Green Energy and Technology

More information about this series at http://www.springer.com/series/8059

Malgorzata K. Gulbinska
Editor

Lithium-ion Battery Materials and Engineering

Current Topics and Problems from the Manufacturing Perspective

Springer

Editor
Malgorzata K. Gulbinska
Pawcatuck, CT
USA

ISSN 1865-3529 ISSN 1865-3537 (electronic)
ISBN 978-1-4471-6547-7 ISBN 978-1-4471-6548-4 (eBook)
DOI 10.1007/978-1-4471-6548-4

Library of Congress Control Number: 2014949339

Springer London Heidelberg New York Dordrecht

Printed on acid-free paper

Springer is part of Springer Science+Business Media (www.springer.com)

Preface I: Introduction

This book covers selected topics in lithium-ion batteries science and technology viewed from a perspective of scientists and engineers who work in an industrial setting. The primary intent of this work was to discuss the current technology, the most pressing challenges, and the potential future developments in an unassuming yet thorough manner. Considering the large volume of publications dedicated to the energy storage in general and lithium-ion batteries in particular, our aim was to provide a "companion" publication that would discuss selected inter-disciplinary aspects of the Li-ion materials and engineering. Modern battery science and engineering often draw inspiration from other science and technology fields, such as physical chemistry, crystallography, electronics, etc. Frequently, the terms and definitions associated with these fields become carried over with an assumption of prior knowledge. Several of these concepts were explained in this publication and/or the sources were given for a more comprehensive reading, as desired by the reader.

Chapter 1 of this book discusses several main cell components: cathodes, anodes, electrolyte solutions, and separators. Typically, the commonly used syntheses methods of these materials are discussed, along with their principal properties and the requirements for successful implementation in high-end cells. The chapter concludes with the forecast of future directions in cell materials' development.

Chapter 2 tackles fundamental aspects of electrochemistry of lithium-ion materials and their relationship with the materials' properties. There is a strong connection between the fundamental electrochemical and physico-chemical properties of materials and the real-life behavior of battery materials. The chapter includes theoretical reasoning as well as the related practical examples.

Chapter 3 discusses optimizing lithium-ion cell materials for various high-end cell designs as well as the particular designs' limitations. Lithium-ion cell components that were individually discussed in the first chapter are considered here as part of the complex system. The relationships between various cell components are illustrated using examples of several successful optimization efforts.

Several high-end lithium-ion applications and the related product requirements are examined in Chap. 4. The high-end batteries are contrasted with their selected mass-produced counterparts. The applications are organized by the environment in which they are used (land, aerospace, space, and sea) with special mention given to the Li-ion batteries for implantable medical devices.

Chapter 5 examines common safety issues commonly encountered in lithium-ion cells and batteries as well as various hazard counteracting measures employed by the lithium-ion manufacturers. The surveyed safety issues pertain to multiple levels of engineering design, from the choice of active materials to full system topography.

Hybrid power sources containing lithium-ion cells or batteries are introduced in Chap. 6. In this chapter, the rationale behind hybrid systems is given, along with the advantages and disadvantages of these devices. Hybrid power systems, such as lithium-ion cell with an ultracapacitor as well as lithium-ion cell with a lithium-air battery are analyzed in this chapter, supported by specific examples of hybrid systems and their demonstrated performance.

Finally, Chap. 7 illustrates the competitive technologies landscape. This chapter assesses some of the most prominent energy storage options competing against lithium-ion batteries. The surveyed systems are organized by the application environment, namely: man portable applications, transportable and stationary applications. A socio-economic commentary is provided for the current and future situation of the selected energy storage market sectors.

This publication can be viewed as an assembly of correlated essays supported by graphs, tables, and pictures, with the common thread being the practical application of several scientific and technological disciplines. The individual chapters were written by professionals who deal with the described challenges in their everyday work. The book is also an illustration on how the high-end technology field benefits from collaborative approach and specialists in various disciplines working together towards the common goal.

Acknowledgments

Malgorzata K. Gulbinska thanks her family and friends for their continued support in work and in life.

Malgorzata K. Gulbinska

Preface II: Historical Notes

Yardney Technical Products, Inc. is a speciality battery manufacturer that figures prominently in this book. This is unsurprising, since the majority of the contributors are either past or present employees of Yardney. The industrial influence was a welcome addition to the book's perspective since the underlying intention of this publication was to place the selected current topics in lithium-ion battery science and technology within the practical context. This historical note presents a few less known facts from the long and rich history of the company and its founder, Michel Yardney.

When Michel N. Yardney came to the United States from Paris in 1940, he had approximately \$500; degrees in Electrochemical Engineering and Electronic Engineering from the University of Caen and Ecole Supérieure d'Electricité; patents of a remote control device he had invented; and a personal motto "Cherche et tu trouveras", "Seek and you shall find". Shortly after his arrival and the start of World War II, he built a working model of the remote control unit, demonstrated the device to the U.S. Navy, and received a \$20,000 prototype development contract. That first small contract was the beginning of the Yardney Electric Corporation, which still holds hundreds of U.S. patents and world-wide patents in the field of electrochemistry. Since that time, Yardney has developed a number of battery systems that include silver zinc, silver cadmium, silver metal hydride, silver hydrogen, magnesium silver chloride, aluminum silver oxide, aluminum air, zinc-air, magnesium-air, lithium-air, cadmium-air, lithium-ion, lithium thionyl chloride, lithium manganese dioxide, and nickel cadmium and nickel hydrogen cells and batteries.

Interestingly, Yardney also developed or supported several commercial products, some that were not associated with batteries at all. These were the "Sea Watch" emergency light, commemorative silver ingots, silver powder, high intensity portable lights, 7.2 hp electric vehicle, electric motorcycle that until very recently held the world speed record for an electric powered motorcycle, a silver doped cellulosic material that was used as an antiseptic article, and other health related items such as Agosan and Eversan first aid ointments and a Como back support system.

Over the years, many of Yardney's battery systems became an integral part of the nation's most advanced spacecraft systems. Some of these products enabled significant technical milestones such as primary power for the astronauts' life support system (Extravehicular Mobility Unit or EMU) that used Yardney's 25 Ah silver zinc battery.

Yardney battery still holds the distinction of going deeper than any manned submersible with the U.S. Navy's Trieste to the bottom of the Marianas Trench using our 1000 and 5000 Ah silver zinc batteries. For many years Yardney's 12 Ah battery was used by virtually every major television network for their portable cameras. Overseas, Yardney's 16 Ah silver cadmium batteries were used by the various prime contractors which included an Italian company for the successful GIOTTO mission that monitored the passage of Halley's Comet past Earth in 1986.

In the early 1990s Yardney capitalized on the potential of lithium-ion battery technology and began the internal R&D program laid the groundwork for transforming the early laboratory work into a lithium-ion battery development program, which eventually led to Yardney becoming a leader in the technology and supporting several space applications including the Mars Rover batteries, multiple satellites, and many manned and unmanned vehicles.

Currently, Yardney is in continuous production of secondary lithium-ion batteries, primary and secondary silver-zinc batteries and primary reserve silver-zinc batteries used on various Department of Defense applications. The primary battery applications include the Navy's Trident II D5 Fleet Ballistic Missile program, the Minuteman III ICBM, and primary power for the MK 21 re-entry vehicle. In 2012, the Trident II missile has achieved 143 successful test launches since 1989—a record unmatched by any other large ballistic missile or space launch vehicle. The most prominent Li-ion batteries made by Yardney have powered the Mars Explorer Rover missions (Spirit, Opportunity, and Curiosity), the USAF B-2 Bomber and Global Hawk aircraft, and the US Navy Advanced SEAL Delivery System (ASDS). One of the future applications for Yardney's Li-ion batteries is NASA's Orion Crew Exploration Vehicle (CEV).

Alexander Karpinski

Contents

Chapter 1
Lithium-ion Cell Materials in Practice

Malgorzata K. Gulbinska

Abstract Modern lithium-ion cells are subject to a wide range of performance requirements, driven by the needs of their end users. Since lithium-ion cells are being used to power devices ranging from household electronics, through power tools to automobiles and the application types vary from commercial to military power sources, there is no universal user profile. The cells are usually classified either as "high energy" or "high power," although the intermediate variants exist. For each distinct cell design option, appropriate selection of cell materials that are "best for the application" is made and, from this point, systematic process of optimization of materials' performance in a cell begins. From the functional standpoint, lithium-ion electrode materials are divided into "active materials" that are capable of reversibly intercalating lithium ions into their structure, "conductive diluents" that assist in electron conduction within the electrode, current collecting foils, as well as binders that assure adhesion to current collectors and cohesion within the electrode. Electrolyte solutions and porous separators are other cell materials that must be properly selected to match the cell design.

1.1 Lithium-ion Cell Components and Materials

Lithium-ion cells (Fig. 1.1) rely on reversible insertion (intercalation[1]) of lithium ions into the structures of cathode and anode active materials, respectively. When active material in an electrode intercalates lithium ions, it becomes electrochemically

[1] Originally, the term *intercalation* was reserved only for the layered materials' host structures; nowadays this term can be seen applied to other types of structures as well (e.g., spinel-structured materials). The term also applies to ions other than Li^+ (e.g., Mg^{2+}).

M.K. Gulbinska (✉)
Yardney Technical Products, Inc., 2000 South County Trail,
East Greenwich, RI 02818-1530, USA
e-mail: mgulbinska@gmail.com

M.K. Gulbinska (ed.), *Lithium-ion Battery Materials and Engineering*,
Green Energy and Technology, DOI 10.1007/978-1-4471-6548-4_1

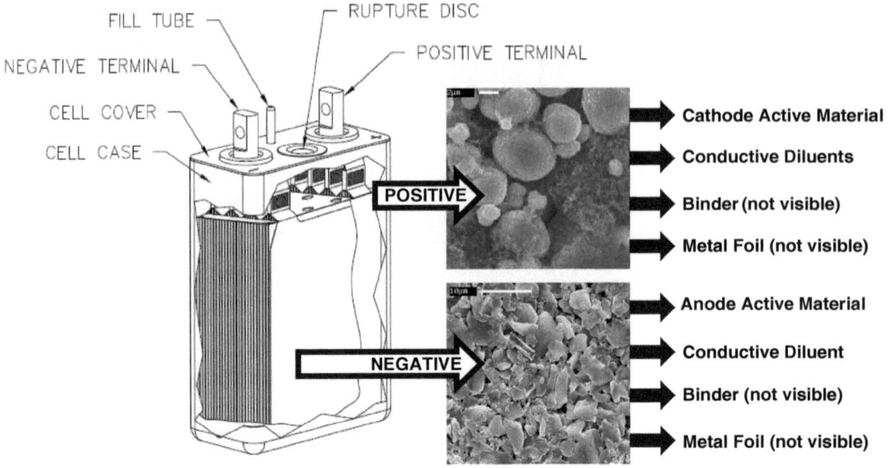

Fig. 1.1 Schematic view of a prismatic 7 Ah cell and electrodes' surfaces (*courtesy of* Yardney Technical Products, Inc.)

Table 1.1 Electrode reactions in a typical lithium-ion cell

Site	Charge	Discharge
Cathode	$LiMO_2 \rightarrow Li_{1-x}MO_2 + x\ Li^+ + x\ e^-$	$Li_{1-x}MO_2 + x\ Li^+ + x\ e^- \rightarrow LiMO_2$
Anode	$C + y\ Li^+ + y\ e^- \rightarrow Li_yC$	$Li_yC \rightarrow C + y\ Li^+ + y\ e^-$
Overall	$LiMO_2 + x/y\ C \rightarrow x/y\ Li_yC + Li_{1-x}MO_2$	$x/y\ Li_yC + Li_{1-x}MO_2 \rightarrow LiMO_2 + x/y\ C$

Note After the first charge of a lithium-ion cell (intercalation of lithium ions into the anode's structure), delithiation of carbon during the subsequent discharges is never 100 % complete; in other words Li_yC does not become fully delithiated to carbon

reduced without undergoing a substantial structural change other than the unit cell volume change. Anode intercalates lithium ions during charge and cathode accepts them during discharge. A complete charge followed by a discharge is referred to as a "cycle." The electrode reactions expressed in general terms and the overall lithium-ion cell reaction are shown in Table 1.1. Active materials constitute about 90 % of the dry mass of both positive and negative electrodes. The balance of each electrode comprises conductive diluents and binder, cast onto the metallic current conducting foil. The function of conductive diluents is to enhance electronic conductivity of active materials and facilitate electronic transport between active material particles and toward the current collecting foil. The need for conductive diluents arises from the limited conductivity of active materials as well as from the fact that the active materials' conductivity varies with the state of lithiation (i.e., fully lithiated lithium cobalt oxide cathode is close to semiconductor in conductivity, whereas the material with the $Li_{0.6}CoO_2$ composition behaves more like a metallic conductor) [1, 2]. The polymeric binder both binds all the constituents of electrode together and attaches them to the current collector.

1.2 Cathode Active Materials

Multiple cathode materials are currently utilized in commercial lithium-ion cells. Table 1.2 lists the most common active cathode materials along with their most important cycling performance characteristics. Derivatives of these cathodes also exist including materials that have been doped with additional ions, surface coated, or otherwise modified. The examples of such derivatives are elaborated in the following sections, along with the rationale for a given modification.

Before being implemented in a commercial cell, a successful cathode candidate material must meet multiple selection criteria. The list below, originally formulated by Whittingham [1] and expanded here, summarizes very well the requirements at the early developmental stages. Further conditions arise during the material processing and implementation into the commercial lithium-ion cells. Moreover, most of the currently used cathode materials fail to entirely satisfy all the expectations.

(1) The cathode material must contain a readily reducible/oxidizable ion, for example, a transition metal, such as cobalt in $LiCoO_2$, which is oxidized and reduced within the Co^{4+}/Co^{3+} redox couple limits. Transition metals are common in this role, mainly due to having more than one stable oxidation state.
(2) The cathode material must react with lithium in a reversible manner.
    (a) This dictates an intercalation-type reaction in which the host structure essentially does not change as lithium is added. Many materials, especially of the layered type, have a preferred voltage range that guarantees

Table 1.2 Selected positive electrode materials

Cathode material	Specific capacity, mAh g^{-1}	Midpoint voltage versus Li at C/20	Comments
$LiCoO_2$	140–155	3.9	Prevalent in portable electronics, cobalt is expensive
$LiNi_{0.8}Co_{0.15}Al_{0.05}O_2$	200	3.73	High-end applications (satellites, etc.), high capacity, safety comparable to $LiCoO_2$
$LiNi_{1-x-y}Co_xMn_yO_2$	140–180	~3.8	One of the prospective automotive li-ion battery cathode candidates, slightly safer than $LiCoO_2$
$LiMn_2O_4$	100–120	4.05	Inexpensive, safer than $LiCoO_2$, poor cycling at elevated temperatures due to manganese (Mn^{2+}) solubility in electrolytes
$LiFePO_4$	160	3.45	Lower energy, safer alternative to $LiNi_{1-x-y}Co_xMn_yO_2$

Note Strictly speaking, discussion of safety in terms of isolated cathode material is not wide enough; other components of a cell should be included in "safety" predictions, such as electrolytes and conductive carbonaceous diluents

Table 1.3 Structural information about selected positive electrode materials

Cathode material	Structure name	Structure symbol; Space group	Isostructural mineral(s)
$LiCoO_2$	Ordered rock salt	$R\bar{3}m$	α-$NaFeO_2$
$LiNi_{0.8}Co_{0.15}Al_{0.05}O_2$			
$LiNi_{1-x-y}Co_xMn_yO_2$			
$LiMn_2O_4$	Spinel	Fd3 m	$MgAl_2O_4$
$LiFePO_4$	Olivine	Pmnb	$(Mg,Fe)_2SiO_4$

the longest host structure stability. The optimal cycling voltage range for maximum cell life is specific for a given material composition and can be predetermined experimentally [3]. The most common cathode materials and their structural properties are given in Table 1.3.

(3) The material should react with lithium with a high free energy of reaction [4].
   (a) High capacity, intercalating preferably one lithium-ion or more per transition metal.
   (b) High voltage, preferably around 4 V (limited by stability of electrolyte, mostly comprised of organic carbonates).
   (c) Combined high voltage and high capacity lead to a high-energy storage.

Text Box 1.1 Thermodynamics and materials' performance

The practical electrochemical parameters (actual cell capacity, cell voltage, etc.) are strongly related to the theoretical thermodynamic calculations and are usually diminished by a certain factor because of the occurrence of various "real-life usage" losses. The most important theoretical properties of battery materials (electrochemical potential of the cell, cell's theoretical capacity, and energy) are derived from thermodynamics of the electrode reactions in lithium-ion cell (Table 1.1). A comprehensive, in-depth discussion of thermodynamics of the processes occurring in a lithium-ion cell can be found elsewhere [4]. Some of the most crucial formulas are listed below.

Standard Gibbs free energy of the intercalation reaction per one mol of reactants (lithium ions):

$$\Delta G_r^{\circ} = z \cdot F \cdot E$$

ΔG_r°—Standard Gibbs free energy of reaction; Joules $\cdot$ mol^{-1}
z—Mobile ion (lithium ion) charge number; equals one for lithium ions (no unit)
F—Faraday constant (96,500 Coulombs per mol of lithium ions)
E—Voltage between the electrodes; expressed in volts

There is a direct relationship between *theoretical -specific capacity* of a cathode material and the Faraday constant (96,500 Coulombs · mol^{-1} = 26.8 Ah mol^{-1}; since 1A = 1C s^{-1}):

$$Q = 1000 \times z \cdot F \cdot M^{1}$$

Q—Specific capacity of the cathode material; expressed in mAh · g^{-1}
M—Molar mass of cathode material, expressed in grams

(4) The material should react with lithium very rapidly both on insertion and removal. This leads to high power density, which is needed to replace the Ni/Cd battery or for batteries that can be recharged using regenerative braking in hybrid electric vehicles.
(5) The material should be a good electronic conductor, preferably a metal.
   (a) This allows for the easy addition or removal of electrons during the electrochemical reaction.
   (b) This allows for reaction at all contact points between the cathode active material and the electrolyte rather than at ternary contact points between the cathode active material, the electrolyte, and the electronic conductor (such as carbon black).
   (c) This minimizes the need for inactive conductive diluents, which take away from the overall energy density.
(6) The material must be stable, i.e., not change structure or otherwise degrade, on either over-discharge or over-charge. This rule is very often broken as maximizing voltage of most materials often degrades their performance and thus shortens their cycle life.
(7) The material must be low cost. The matter of cost is somewhat relative; in high-end applications such as satellite batteries, cost is of secondary importance and factors such as high energy or long cycle life take precedence.
(8) The material should be environmentally benign. This criterion is typically difficult to fulfill as many transition metals, such as cobalt and nickel, are toxic to various degrees. The matter becomes even more complicated with introduction of nanoparticles that pose their own environmental challenges. A more detailed review of nanoparticles-based materials and the technical and environmental aspects of their use will be done in the latter chapter of this book.

1.2.1 Lithium Cobalt Oxide

Lithium cobalt oxide ($LiCoO_2$) has been used as cathode material in lithium-ion cells for portable applications (computers, cell phones, and cameras) since the introduction of the first commercial lithium-ion cell by Sony [5]. A decade

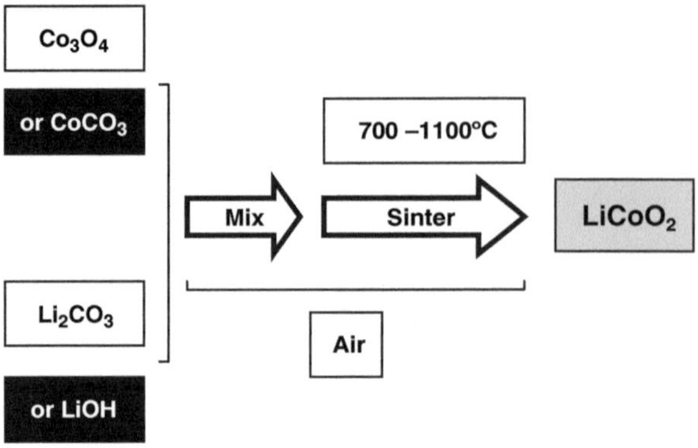

Fig. 1.2 Solid state syntheses routes of $LiCoO_2$ cathode material

before the commercial application, $LiCoO_2$ was demonstrated in a 4 V prototype rechargeable cell by Goodenough and Mizushima [6, 7] against lithium metal as the negative electrode.

Despite the relatively high cost and scarcity of cobalt, $LiCoO_2$ is still the most commonly used commercial cathode for portable electronics due to multiple performance advantages offered by this material: flat discharge curve profile at relatively high voltage versus lithium (3.9 V), high coulombic efficiency (>92 % [8] charge vs. discharge) during the first cycle and good thermal stability [9, 10]. In addition to the advantageous cell performance characteristics, $LiCoO_2$ is relatively easy to synthesize via solid state synthesis route (Fig. 1.2) and the synthetic method itself is relatively inexpensive.

During the syntheses of $LiCoO_2$, the growth of cathode particles is controlled by the firing temperature and time, and by the amount of added lithium source (Li_2CO_3 or LiOH). A few atomic percent (usually 5–6 at.%) of excess Li per Co atom favors large (about 15 μm) $LiCoO_2$ crystal growth which is desired due to lower reactivity of cathode particles with organic electrolyte. However, part of the synthetic excess lithium enters the $LiCoO_2$ structure and lowers capacity of the final $LiCoO_2$ product, because some Co^{3+} which is redox active is replaced by inactive Li^+ [11].

Several key aspects of the original commercial lithium-ion cells have been reviewed by Nishi, who discussed the need for large particle size, 15–20 μm, to increase safety and the intentional incorporation of lithium carbonate into the cathode to provide a safety valve. The excess lithium carbonate decomposes, releasing carbon dioxide if the charging exceeds 4.8 V, which breaks the electrical circuit and stops the cell's operation.

The average capacity of the $LiCoO_2$ cell is relatively low at about 140–155 mAh g^{-1} because only around 0.5 Li/Co can be reversibly cycled without causing severe cell capacity loss due to phase change in the $LiCoO_2$ structure [12]. $LiCoO_2$ has the α-$NaFeO_2$ structure (space group $R\bar{3}m$, Table 1.3) with the oxygen atoms arranged in a cubic close-packed array, as seen in Fig. 1.3. Structure-property relationship, such as cathode crystal structure changes' correlation to the cell's

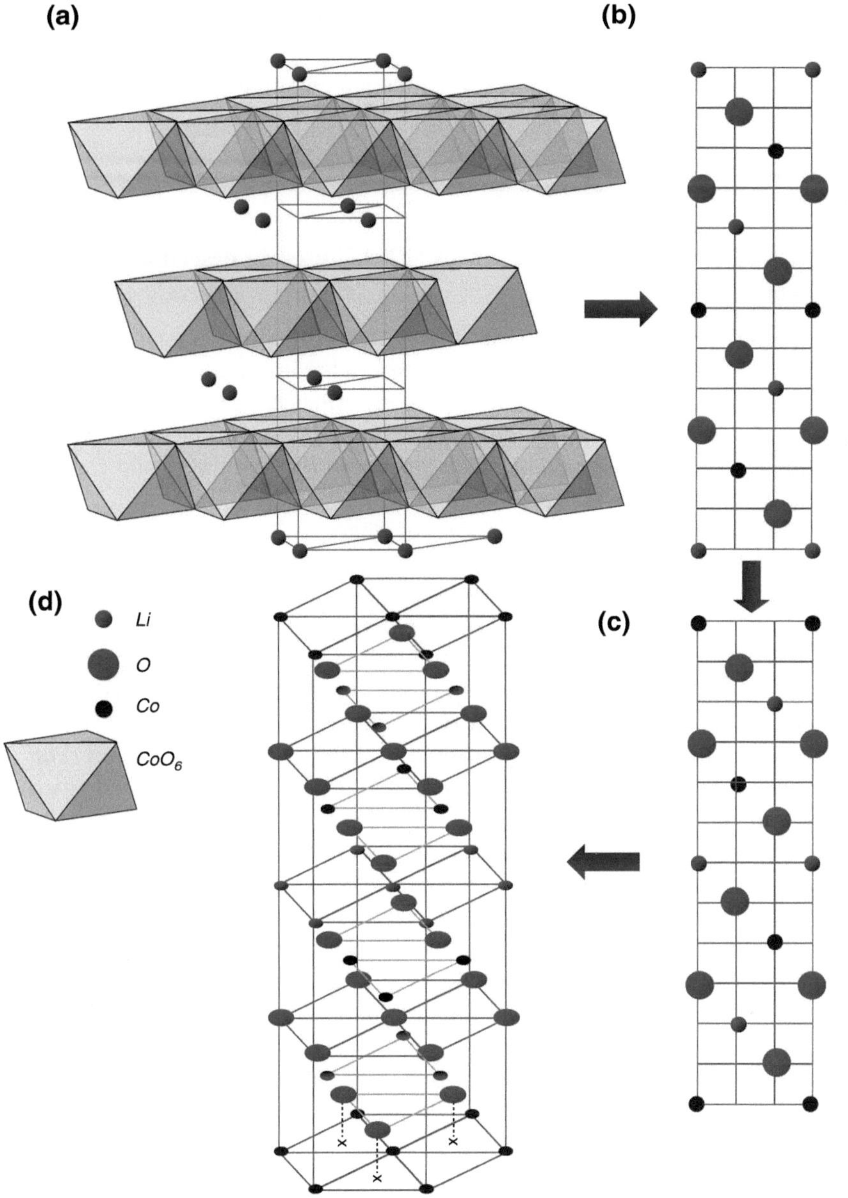

Fig. 1.3 $LiMO_2$ structure (M = Co or Ni) (**a**); schematic stacking of $LiMO_2$ structure building blocks, including oxygen layers arranged in the …(ABCABC)… manner, otherwise known as an O3 array (**b**); the schematic from (**b**) translated along the c-axis (**c**) and; hexagonal unit cell representation of the ordered rock salt $LiMO_2$ structure (**d**); *Note* atoms in schematics (**b**) and (**c**) are projected onto one plane despite being actually positioned in several parallel planes

capacity fade, is very strong in case of $LiCoO_2$ cathode and remains important in other cathode chemistries. Therefore, a small section of this chapter is dedicated to explaining selected crystallographic terms that are commonly found in the battery materials-related literature (Text Box 1.2).

Text Box 1.2 Crystallographic terms [13, 14] used in describing battery materials

Vast majority of battery materials possess a well-defined crystal structure. A basic understanding of crystal symmetry helps in relating the materials' structural properties with their cell performance. Great self-teaching resources can be found on the International Union of Crystallography website [13] and elsewhere [14]. The most common symmetry and crystallography-related terms are listed below.

Point group is a mathematical term and denotes a set of symmetry operations (mirror planes, screw axes, etc.) that can be done on a finite entity (for example, a geometrical shape). This set of operations when applied to the entity, will leave at least one point unchanged in terms of its position.

Space group term refers to the symmetry of a crystal structure (an infinite entity) and, similarly to the point group, also denotes a set of symmetry operations but with added translation operations, which allows to describe the entire crystal structure as the repetition of the smallest unique unit (*unit cell*). The rules that govern the definition and use of point groups and space groups are part of the *group theory*.

Symbols of point groups and space groups are based on two notation systems: *Schoenflies* notation and *Herman–Mauguin* notation.

Wyckoff positions classify the positions of atoms in a crystal structure and describe the "outcome" of all the symmetry operations allowed for a given space group on the atoms contained in the unit cell. The Wyckoff position symbols consist of a number denoting the multiplicity of the position and a letter label (for example, oxygen atoms in the $LiCoO_2$ structure occupy the Wyckoff positions denoted as 6c). The multiplicity of the position informs on how many atoms are generated by symmetry operation on a single atom at that position. The letters are label only and are assigned alphabetically, from the atom residing on the symmetry element with the highest symmetry (bottom of the table, starts from letter "a") to the position with lowest symmetry (top). The uppermost Wyckoff position, called a general Wyckoff position, corresponds to an atom at an arbitrary position that never resides upon any symmetry elements. Wyckoff positions corresponding to each of the 230 possible crystallographic space groups are listed in the *International Tables for Crystallography*.

Previously discussed structural change that impacts the capacity loss in $LiCoO_2$ material is one of the possible causes for cathode performance degradation over time. Another potential cause of capacity loss in lithium cobalt oxides is related to cathode-electrolyte interactions. The electrode's surface interaction with organic carbonate-based electrolytes is a known problem in many different cathode types. Surface coatings, such as phosphates [15, 16] or surface and/or lattice doping (e.g., with Al, Mg, or Ti) have been known to improve cathodes' cycle life and capacity fade at elevated temperatures (≥50 °C). The mechanism of the surface coating-related protection of $LiCoO_2$ cathode was explained by minimizing the reactivity of Co^{4+} on charge with hydrofluoric acid (HF) traces in the electrolyte coming from the interaction of moisture with the electrolyte salt lithium hexafluorophosphate ($LiPF_6$). Removing the source of the HF (the $LiPF_6$ salt) was also found to diminish the capacity loss in $LiCoO_2$, where replacing the $LiPF_6$ salt by lithium bis(oxalato)borate (LiBOB) or by completely drying the $LiCoO_2$ by heating to over 550 °C, improved the capacity retention at 180 mAh g^{-1} at a 4.5 V cutoff. Surface treatments and surface or lattice doping are common practices in attempts to improve cathodes' life, elevated temperature performance as well as safety characteristics in cells.

Indirectly related to the structural properties is the diffusion coefficient of lithium ions in $LiCoO_2$, which is relatively high at 5×10^{-9} cm^2 s^{-1}. This high diffusivity value is consistent with the ability to cycle $LiCoO_2$ cathodes at current rates up to 4 mA cm^{-2}. However, the conductivity of $LiCoO_2$ radically changes with the level of lithiation, from 2 to 4 orders of magnitude at 25 °C up to six orders of magnitude at lower temperatures. The changes of conductivity with lithiation levels combined with the modest practical capacity of $LiCoO_2$ and with the relatively high cost of cobalt provided the incentives for development of other cathodes with layered structure, such as $LiNi_{0.8}Co_{0.2}O_2$ and $LiNi_{0.8}Co_{0.15}Al_{0.05}O_2$ and the family of $LiNi_{1\text{-}x\text{-}y}Co_xMn_yO_2$ cathodes.

Text Box 1.3 The ways of expressing charge and discharge rates

Expressing the charge or discharge current rates does not give much information on how these current rates are related to the size of battery or cell. There are two dominant ways of normalizing the charge or discharge current rates in lithium-ion cells:

- Current rate (A or mA) per surface area of electrode; commonly expressed in mA · cm^{2}.
- C-rate; dimensionless; the C-rate is a measure of the rate at which a battery is discharged, relative to its maximum capacity. A 1C rate means that the discharge current will discharge the entire battery in 1 h. For a battery with a capacity of 100 Ah, this equates to a discharge current of 100 A.

1.2.2 *Lithium Nickel Oxide Derivatives: $LiNi_{0.8}Co_{0.2}O_2$ and $LiNi_{0.8}Co_{0.15}Al_{0.05}O_2$*

Lithium nickel oxide, $LiNiO_2$, is isostructural with $LiCoO_2$; in both oxides crystal structure consists of layers of transition metal cations separated from Li layers by oxygen (Fig. 1.3). Despite $LiNiO_2$ having higher specific energy, being made from the more abundant and less expensive raw material precursors, and nickel being less toxic than cobalt, $LiNiO_2$ has not been pursued in the pure state as a lithium-ion cell cathode, due to potential problems with this cathode material's cyclability safety, and thermal stability in lithium-ion cells. The multiple problems associated with $LiNiO_2$ cathodes originate predominantly from their structural instability.

First, the formation of nonstoichiometric $Li_{1-\delta}Ni_{1+\delta}O_2$ occurs preferentially instead of the formation of stoichiometric $LiNiO_2$, especially at high temperatures during the solid state syntheses (>1100 K or 726 °C). The highly mobile Ni^{2+} ions migrate to the Li sites and disrupt lithium diffusion, thereby reducing the reversible electrochemical capacity, as well as the lithium diffusion coefficient and the power capability of the $LiNiO_2$ electrode. The formation of oxygen and lithium vacancies was found to be thermodynamically favorable in $LiNiO_2$ at high temperatures, which in addition to high mobility of Ni^{2+} ions provides multiple routes for the formation of defected, nonstoichiometric lithiated nickel oxide structures [17].

Second, once the nonstoichiometric $Li_{1-\delta}Ni_{1+\delta}O_2$ compounds with low lithium contents are formed, they appear to be unstable due to the high effective equilibrium oxygen partial pressure. The high oxygen partial pressure points to the tendency of oxygen to leave $LiNiO_2$ crystal lattice, which would be aggravated at highly charged state of cathode (~4.4 V and above). The evolution of oxygen in the presence of organic solvents from electrolyte renders $LiNiO_2$-based cells inherently unstable and unsafe.

Preparing $LiNiO_2$ with an excess amount of Li was found to be one of the methods to produce stoichiometric cathode material. Another, even more effective method for stabilizing the lithium nickel oxide structure was the introduction of cobalt and thus formation of the lithiated nickel cobalt oxide derivatives of $LiNiO_2$ material. These findings led to laboratory development and commercial production of various derivatives of lithium nickel oxide, summarized in Table 1.4.

1.2.3 *The Family of $LiNi_{1-x-y}Co_xMn_yO_2$ Materials*

Unlike cobalt and nickel, manganese does not form a stable, pure $LiMnO_2$ phase and forms the spinel structure instead, named after the mineral of spinel (Table 1.3) with the composition of $Li_{0.5}MnO_2$, commonly expressed as $LiMn_2O_4$. Manganese-based cathode materials are attractive mainly due to their low cost compared to cobalt and nickel. A wide variety of the lithiated manganese oxides with multiple structures and compositions (Li:Mn:O ratios) exist and many of these compounds can reversibly intercalate lithium, but amongst these materials, spinel materials and the layered derivatives are the most mature, to the point of commercial availability. Spinel-structured materials will be discussed separately in the next part of this chapter.

Table 1.4 Selected $LiNiO_2$ derivatives and rationale for their modification

Cathode material and sources	Rationale for modification	The disadvantage of the modified material
$LiNi_{0.8}Co_{0.2}O_2$ Commercial	Cobalt assists in ordering the layered structure and suppresses the nickel migration into the lithium layer;	Cobalt-containing derivatives suffer from the drastic conductivity changes with lithiation, similar to those observed in $LiCoO_2$ materials
	Cobalt-substituted $LiNiO_2$ derivatives are less likely to lose oxygen than the pure nickel oxide, which makes them safer than pure $LiNiO_2$	
$LiNi_{0.8}Co_{0.15}Al_{0.05}O_2$ Commercial	Aluminum, being redox inactive, prevents the complete removal of all the lithium, thus additionally stabilizing the structure and preventing any phase changes that might occur at very low or zero lithium content	The composition-induced changes in conductivity of cathode materials demand that a conductive diluent be added to the cathode active material, which reduces both the energy storage and the power capabilities
$LiNi_{1-x}M_xO_2$ $M = Mg$, Sr, or B Developmental	The rationale is similar to Al introduction	Additionally, the electronic conductivity is adversely impacted when *any* "foreign" atom is introduced into the $LiNiO_2$ structure

$LiMO_2$ compounds containing manganese require the presence of cobalt and nickel to possess the stable layered structure. The most popular member of this family of mixed metal oxide compounds is the 333 (or 111) NCM (nickel, cobalt, manganese) material, named so because of having the formula of $LiNi_{0.33}Mn_{0.33}Co_{0.33}O_2$. The NCM 111 cathode can deliver 150 mAh g^{-1} when cycled between 2.5 and 4.2 V. Raising the charge cutoff potential would increase the capacity significantly (about 220 mAh g^{-1} at 5.0 V) but at the cost of rapid capacity fade.

Text Box 1.4 Solid solutions and their voltage against time profiles

The dramatically shortened cell life as a consequence of the extended voltage range is common for many layered materials, which are usually solid solutions of mixed metal oxides and have sloped voltage against time profiles. Voltage extension is typically done in order to extract extra capacity (and energy) out of the cathode material. When dQ/dV (Q—capacity, V—voltage) derivative is taken for solid solution-type materials, one can clearly see that capacity is delivered over rather large voltage range (interval) and the derivative has a shape of a diffused peak (single or complex, depending on the exact cathode composition); unlike in multiphase cathodes, where voltage versus time profile is very flat and the dQ/dV derivative is usually a sharp, narrow peak (or peaks) occurring at well-defined voltage value.

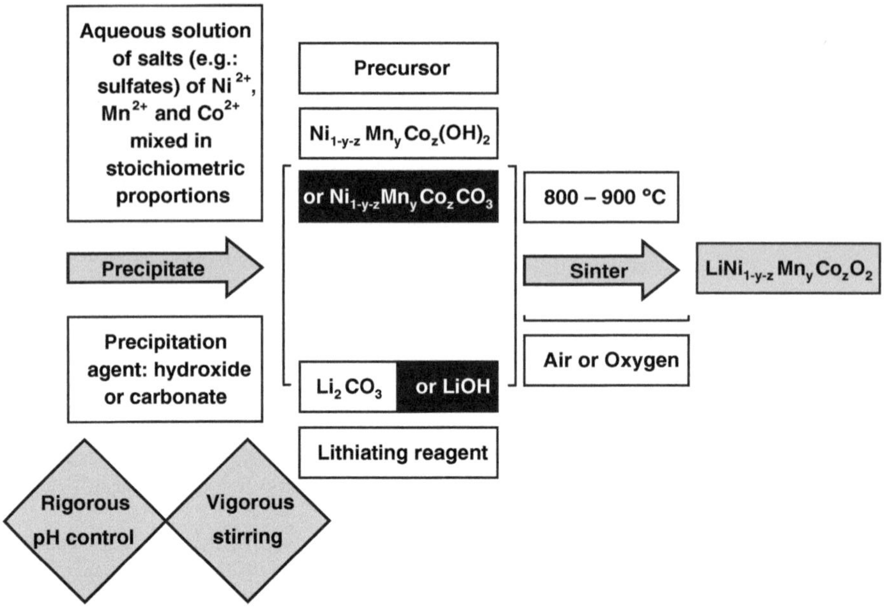

Fig. 1.4 Multistep syntheses routes of $LiNi_{1-y-z}Mn_yCo_zO_2$ materials

Unlike $LiCoO_2$ cathode material that is synthesized via a relatively simple solid state process, the $LiNi_{1-y-z}Mn_yCo_zO_2$ compounds are usually prepared using a multistep process based on the modified mixed-hydroxide approach. The first step is precipitation of a mixed metal hydroxide, $Ni_{1-y-z}Mn_yCo_z(OH)_2$, which is then reacted with a lithium salt in air or oxygen, with the optimum temperature of 800–900 °C. The mixed metal hydroxide precipitation (alternatively, mixed metal carbonate) is a crucial step, requiring rigorous parameters control (Fig. 1.4).

In all NCM materials, regardless of the exact chemical composition, nickel is the electrochemically active metal that is oxidized when lithium is removed and reduced when lithium is reintroduced back into the cathode structure. Nickel oxidation proceeds in two redox processes: Ni^{2+}/Ni^{3+} and Ni^{3+}/Ni^{4+} throughout the entire cycling range of the cathode material. Cobalt redox couple Co^{3+}/Co^{4+} seems to be active predominantly at high potential of 4.54 V [18]. In the discharged state (fully lithiated cathode), nickel is predominantly divalent, cobalt is trivalent and manganese is tetravalent, which suggests the spectator role of manganese in NCM materials. The role of cobalt in NCM cathodes is mostly keeping nickel out of the lithium layer and thus improving the rate capability of material, enabling the unobstructed movement of lithium-ions through cathode structure. The good rate capability of NCM materials is observed despite their modest electronic conductivity that varies with the synthesis method [1, 19, 20].

Manganese, in addition to lowering the cost of raw materials for NCM cathodes, impacts the safety (specifically, structural and thermal stability) of these materials [21]. In X-ray diffraction studies of various layered cathodes done at temperatures varying from 25 to 900 °C, manganese-containing layered cathodes have shown enhanced structural stability demonstrated by their α-$NaFeO_2$ structure retention. The structural stability was explained by stronger covalent bonding compared to the LNC materials, induced by the presence of Mn^{4+} ions in the NCM structure. The comparative differential scanning calorimetry (DSC) studies have shown the same onset temperatures for the electrolyte combustion in the presence of a cathode material for NCM and LNC cathodes but a notably lower heat flow with NCM materials [21]. These structural and thermal stability improvements should not be overstated as the safety of a lithium-ion cell depends on many factors, such as anode chemistry, electrolyte composition, cell shape and size, battery design, etc.

1.2.4 Lithium Manganese Spinel and Derivatives

Lithium manganese oxide spinel belongs to a different structure class (Table 1.3, Fig. 1.5) from the previously discussed layered $LiMO_2$ materials (M = Co or Ni). The molecular formula of lithium manganese oxide spinel, commonly expressed as $LiMn_2O_4$ is based on atomic ratios of 1:2:4 (Li:Mn:O).

Cost-effectiveness, low toxicity, and good stability at highly charged (strongly delithiated) state are the most attractive features of $LiMn_2O_4$. In this case, the term "stability" refers to the lack of the propensity in spinel to release oxygen from the material, while the advanced delithiation occurs. The lack of oxygen release stands in contrast to the layered materials, which are more unstable when overcharged. The stability of lithium manganese oxide spinel has a structural origin (Fig. 1.5) and is related to stronger M–O (M = Mn) bonds in spinel than those observed in layered $LiMO_2$ materials. The modest specific capacity of $LiMn_2O_4$ (100–120 mAh g^{-1}) is partially compensated by a high midpoint discharge potential (4.05 V vs. lithium) when approximating the specific energy delivered by $LiMn_2O_4$. Approximated specific energy of a cathode (Wh kg^{-1}) equals to the maximum specific capacity (Ah kg^{-1}) multiplied by midpoint voltage for this cathode).

The initial development stages of lithium manganese spinels were plagued by problems with their cycle life, especially at elevated temperatures (≥55 °C). The capacity fading of $LiMn_2O_4$ has been explained by the following two major factors: (1) the slow dissolution of manganese (Mn^{2+}) in organic electrolyte solutions containing $LiPF_6$ salt and (2) Jahn–Teller distortion in deeply discharged state due to the prevalence of Mn^{3+} in $Li_xMn_2O_4$ ($x \geq 1$). The increased concentration of Mn^{3+} ions in the host structure, followed by the increase in occurrences of Jahn-Teller distortion on a molecular level, introduces a cubic to tetragonal distortion of the spinel's crystal structure and causes the damage to cathode's structural integrity upon repeated cycling, manifested as capacity fade over time. Careful selection of the cycling voltage range, i.e., avoiding over-discharge of $Li_xMn_2O_4$ spinel cathodes significantly decreases this cause for capacity fade.

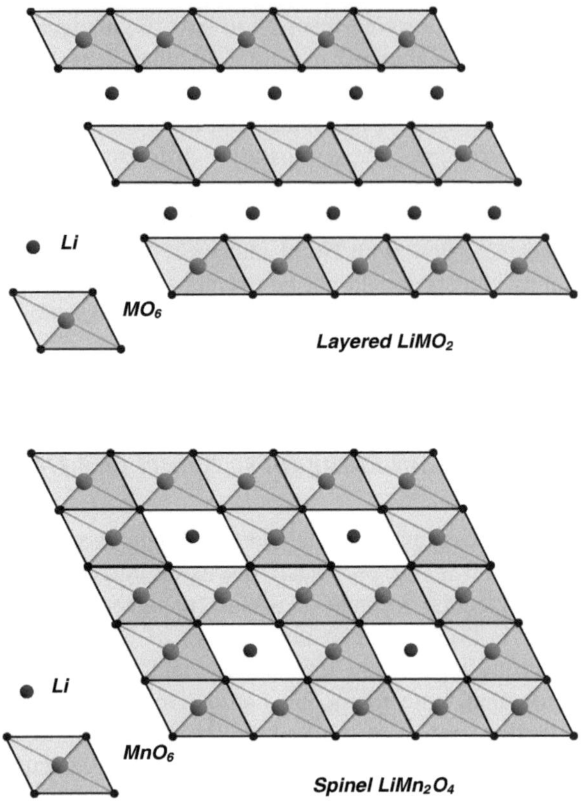

Fig. 1.5 Layered $LiMO_2$ structure (M = Co, Ni and/or Mn) and spinel structure of $LiMn_2O_4$ (*bottom*)

The manganese dissolution in electrolyte caused by HF-yielding hydrolysis of $LiPF_6$ salt exposed to traces of moisture is believed to be the main cause for the loss of cycle life in $Li_xMn_2O_4$ spinel-based lithium-ion cells. The dissolution of lithiated manganese oxide cathodes is similar to lithium cobalt oxide cathode dissolution in $LiPF_6$–containing electrolytes, except that in case of lithium cobalt oxide, the dissolution process is aided by the presence of the oxidative Co^{4+} (charged cathode) and in $Li_xMn_2O_4$ ($x \geq 1$), the most soluble manganese species (Mn^{2+}) come from the disproportionate reactions of Mn^{3+} ($2Mn^{3+} \rightarrow Mn^{2+} + Mn^{4+}$) that is more populous in the highly discharged cathode [22].

Text Box 1.5 Jahn–Teller distortion applied to $LiMn_2O_4$ spinel

Jahn–Teller theorem relates the electronic configuration with the resulting change in geometry of a metal complex (e.g., MnO_6). Specifically, in the case of Mn^{3+} ion (d^4 configuration), the four d electrons are "unevenly" distributed on the three degenerate (energetically indistinguishable) molecular orbitals of the octahedral MnO_6 complex, which is the "building block" of the $Li_xMn_2O_4$ spinel.

Following the Jahn-Teller theorem, the *z*-axis elongates and thus, part of the degeneration is relieved. Only two molecular orbitals are now energetically identical and they contain two pairs of electrons. In $Li_xMn_2O_4$ spinel ($x \geq 1$), the molecular geometry change extends onto the crystal geometry change from cubic (all unit cell dimensions are equal) to tetragonal unit cell (the unit cell is elongated in a *z*-direction).

The Jahn–Teller distortion in $Li_xMn_2O_4$ spinels has the most pronounced effect when the average oxidation state of manganese is lower than 3.5 ($x > 1$), which is mostly enabled in half cells (anode is Li/Li^+) [22] or within a battery, in a malfunctioning (e.g., possessing uncharacteristically low capacity) cell. The electrochemical over-lithiation (over-discharge) of the cathode requires an additional source of lithium in the cell. This is because most of full lithium-ion cells (typically containing graphitic anodes) start their life in the fully discharged state, with a fully lithiated cathode and completely delithiated anode.

The three pathways to counteracting the manganese dissolution in electrolyte are: (1) partial substitution of manganese in cathode material with heteroatoms, (2) surface modifications of the cathode, and (3) electrolyte modifications.

Partial substitution of manganese ions by trivalent or divalent cations like Ni, Co, Cr, Al, and Mg decreases the Mn^{3+} content in favor of Mn^{4+} by virtue of utilizing the charge balance of $LiMn_2O_4$ and thus improves the stability of cathode in the discharged state. Alternatively, coating the surface of the spinel particles with materials, such as zirconium dioxide or $AlPO_4$ is another way of improving the cycle life of $LiMn_2O_4$ cathodes. The rationale behind using zirconium dioxide or $AlPO_4$ is that they are believed to act as scavengers of HF. Another method of $LiMn_2O_4$ cycle life stabilization relies on replacing the fluoride-containing $LiPF_6$ salt with LiBOB, which, similarly to the strategy previously used with $LiCoO_2$ cathodes, eliminates the trace HF amounts from electrolyte.

Sometimes, an introduction of excess lithium is done in lithium manganese spinels (typically, less or equal to 6 atomic % excess lithium). The excess lithium improves the cathode's stability in the charged state and the cycling stability of the material. The presence of extra lithium in the spinel lattice also "pins" lithium ions into their sites in the crystal lattice and prevents the potential migration of mobile lithium ions into manganese lattice sites.

1.2.5 Lithium Iron Phosphate

Lithium iron phosphate ($LiFePO_4$) was first reported in the English technical literature by Padhi et al. [23] from John B. Goodenough's group. This cathode material is very attractive for commercial battery applications because iron is abundant in nature, relatively inexpensive and has low toxicity, especially when compared to nickel or cobalt.

The midpoint discharge potential of $LiFePO_4$ versus lithium is relatively low at 3.45 V at C/20 discharge, compared to about 3.8 V for $LiNi_{1-x-y}Co_xMn_yO_2$ cathodes and 4.05 V for $LiMn_2O_4$ [24]. However, the capacity of $LiMn_2O_4$ is about 40–60 mAh g^{-1} lower than that of $LiFePO_4$ (>160 mAh g^{-1}) and the $LiNi_{1-x-y}Co_xMn_yO_2$ cathodes (140–180 mAh g^{-1}) exhibit sloping voltage profiles, unlike the flat voltage profile of $LiFePO_4$. As stated previously, the maximum specific capacity multiplied by midpoint values can be used to approximate the specific energy of a material, but the entire discharge voltage versus time curve needs to be taken into consideration when more than approximated calculations of specific energy are done. The flat voltage versus time profile combined with very good capacity retention after several hundred cycles further increases the attractiveness of $LiFePO_4$ material as the lithium-ion cell cathode.

Lithium iron phosphate crystallizes in the olivine phase (Pmnb). The ion arrangement in $LiFePO_4$ crystal structure differs significantly from that in the layered and spinel cathodes, described in previous sections of this chapter. The $LiFePO_4$ structure is composed of two types of building blocks: PO_4 tetrahedra and FeO_6 octahedra. Each octahedron shares edge with one tetrahedron in the *c*-direction and two corners with two other tetrahedrons within the *ab*-plane. There is no continuous network of FeO_6 or PO_4 moieties and only one-dimensional channels are available for lithium-ion mobility. In consequence of such structure, $LiFePO_4$ possesses low intrinsic electronic conductivity (10^{-7}–10^{-9} s cm^{-1}) and the low diffusion coefficient of Li^+ (10^{-11}–10^{-13} cm^2 s^{-1}).

On the other hand, the entirely delithiated, stable $FePO_4$ compound exists and also has olivine structure [25]. The existence of the robust delithiated form is a good prognostic in terms of the relative over-charge resilience of $LiFePO_4$, which has been demonstrated in practice [26]. Another unique aspect of the $LiFePO_4$ is that the oxygen atoms are strongly bound within the PO_4 groups. The strong P–O bond renders the oxygen atoms less prone to be released on overcharge of the cathode and to react with the organic electrolyte.

Several synthetic strategies were employed by various groups to address the low electronic and ionic conductivity of $LiFePO_4$. The most common syntheses methods are summarized in Table 1.5 and the synthetic strategies to improve ionic and electronic conductivity of $LiFePO_4$ are summarized and Table 1.6.

Low-temperature routes such as hydrothermal syntheses (Table 1.5) tend to yield the desired small particle sizes but care must be taken when optimizing the syntheses conditions such as time and temperature. Insufficient synthesis time will yield $LiFePO_4$ that apparently has the correct olivine structure as evidenced by the routine X-ray diffraction investigation but upon a closer inspection there are a few percent iron atoms in the lithium sites that block diffusion of the lithium ions and have detrimental effect on electrochemical performance of such material. Firing the hydrothermal material to 700 °C was found to resolve this disorder. Alternatively, the hydrothermal approach can be much improved to produce

Table 1.5 Common $LiFePO_4$ syntheses methods

Synthesis method	Temperature, °C	Notes
Solid state	650–700	Simple, easily industrialized method
		Lengthy, cumbersome process
		Yields nonuniform particles
Hydrothermal	120	Simple process
		Low energy consumption
		Yields relatively small particles
		Platelet morphology of $LiFePO_4$ synthesized by this method is not the most desirable for batteries (spherical particles are currently the most desired)
Sol-gel	300–500	Excellent control of the product properties (composition, particle size, and uniformity)
		Low syntheses temperatures
		Simple equipment
		Difficult industrialization (scale-up)
Microwave-assisted heating	<300	Rapid reaction times (<30 min)
		Uniform heating
		Challenging scale-up
		Nanorod morphology reported by some groups (Manthiram); this morphology could be advantageous (good for high rate) if nanorods are oriented
Other methods	Carbothermal reduction	
	Spray pyrolysis	
	Mechanical activation method (ball milling)	

materials with excellent electrochemical behavior by adding reducing agents such as ascorbic acid to the reaction mixture. The reducing agents prevent surface ferric films formation that would otherwise decrease the material's conductivity.

1.3 Anode Active Materials

Theoretically, lithium metal has the highest possible specific capacity of all possible anodes in lithium-ion cells. In practice, during repeated cycling, lithium is unevenly replated upon reduction from Li^+ back to Li metal, which leads to the formation of high-surface, reactive lithium dendrites. These dendrites tend to grow into the polymer cell separator and may cause electrical short circuit and the resulting safety hazard in a lithium-ion cell.

Currently, the most common commercial anodes are carbons that possess relatively low surface area and intercalate lithium ions into their layered structure without forming metallic lithium under standard operating voltage, temperature, and current rate conditions. Several types of carbon that are commercially available as anodes for lithium-ion batteries are summarized in Fig. 1.6.

Table 1.6 $LiFePO_4$ syntheses optimization strategies

Synthesis strategy	Notes
Lowering particle size	Yamada et al. [27] proposed decreasing the particle size of $LiFePO_4$ as an effective way to overcome the diffusion limitation problems and the resulting insufficient ionic conductivity of $LiFePO_4$
Carbon addition/coatings	Samples made from reagent-grade carbon-containing materials have a conductivity of around 10^{-5}–10^{-6} s cm^{-1}
Addition of reducing agents during syntheses	Ravet et al. [28] first reported that coating with about 1 wt% carbon resulted in improved conductivity and electrochemical performance of $LiFePO_4$ Addition of reducing agents prevents the undesired iron oxidation to Fe^{3+} during synthesis (in fully lithiated $LiFePO_4$, Fe^{2+} is desired) and in most cases produces a thin coating of conductive carbon [29]
Noncarbonaceous coatings and additives	Metallic (e.g., silver) [30] or thin, conductive inorganic coating (Fe_2P or FeP/Fe_2P mixture) performs similar function to carbon coatings. [31]
Doping	Chung et al. [32] showed increase in $LiFePO_4$ conductivity by 8 orders of magnitude upon "doping" with elements such as niobium The conductivity increase was found [33] to be actually related to the formation of a highly conductive iron phosphide, Fe_2P, surface film, formed at high temperature in the presence of a reducing (carbon)

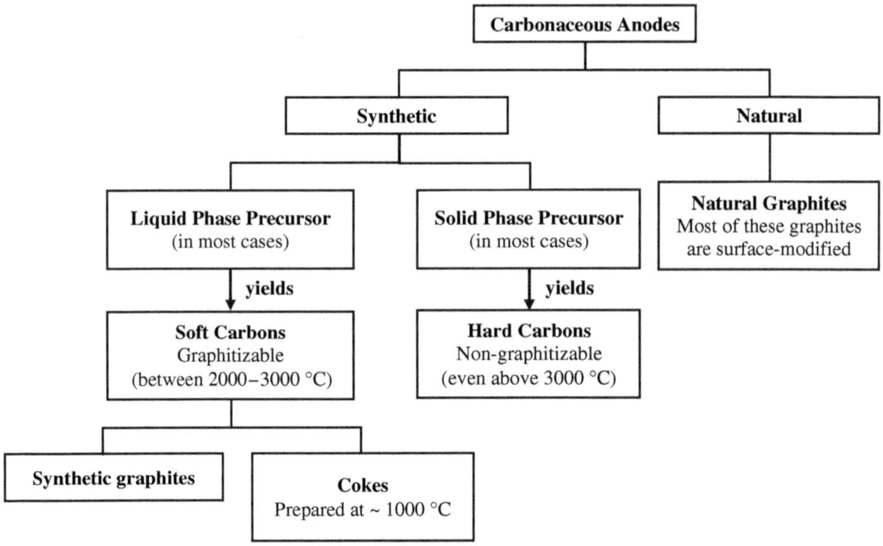

Fig. 1.6 The origin and classification of carbonaceous anodes for lithium-ion batteries

The common basic building unit in carbons is a graphene sheet, which constitutes of a planar arrangement of multiple C_6-hexagons joint by their edges in a honeycomb fashion. In graphites, graphene layers are stacked in an organized manner: the ...ABABAB... stacking occurs in hexagonal (2H) graphite and the ...ABCABC... stacking is observed in rhombohedral (3R) graphite. Majority of the carbonaceous materials contain domains of both 2H and 3R arrangements as well as randomly stacked fractions. In soft carbons, upon increasing of the graphitization temperature above 2000 °C, the material becomes more ordered, which is desirable in terms of obtaining the specific capacity closest to the theoretical value of 372 mAh g^{-1}. The theoretical capacity of graphite is based on the theoretical intercalation of one mole of lithium ions per one C_6-unit.

When lithium is intercalated into graphite, the ...ABABAB... stacking becomes the ...AAAAAA... arrangement and in highly ordered graphites, three very distinct voltage plateaus are observed with the progressing lithiation. This is called the lithium staging and is caused by the fact that with the increasing degree of lithiation, the distribution of lithium is localized in clusters. Moreover, due to repulsion between clusters, alternate layers of graphite are filled consecutively with lithium ions until they are entirely occupied, at which point the remaining layers are filled. The lithiation progresses from the unlithiated C_6 form, through partially lithiated Stages 3 and 2, to the fully lithiated Stage 1 (LiC_6).

Selected commercial carbons from various manufacturers are summarized in Table 1.7 along with their physical and electrochemical properties. Several trends are ought to be considered when selecting the anode material:

(1) The specific surface area of the powders (expressed in $m^2\ g^{-1}$; usually measured by Brunauer-Emmett-Teller method of nitrogen adsorption—BET) is inversely proportional to their median particle diameter (d_{50}).
(2) Usually, with smaller particles, high discharge rate capability of anode is improved but the possible trade-off is experienced in the irreversible capacity loss (ICL) during the first charge/discharge cycle. The ICL on the anode's side is mostly related to the reductive surface reaction between electrolyte and anode's surface, otherwise known as solid-electrolyte interphase (SEI) formation.
(3) Surface treatments (coatings) on natural graphites are usually directed at reducing their specific surface area and minimizing their interactions with electrolyte solutions.
(4) Graphitic materials are susceptible to co-intercalation (along with Li^+ ions) of propylene carbonate (PC), one of the vital constituents of low-temperature electrolytes. Surface treatments of graphitic materials are one of the strategies aimed at alleviating of the PC co-intercalation issue; this work is still ongoing.
(5) Thermal purification of graphites usually has twofold purpose: reducing the ash (inorganic oxides, such as SiO_2, Al_2O_3, Fe_2O_3, etc.) content and removing most of the structural disorder in the material (e.g., turbostratic disorder where the consecutive graphene layers are planar and parallel with respect to each other but rotated.

Table 1.7 Selected carbonaceous anodes for lithium-ion batteries [34–38]

Manufacturer	Grade	Type	BET surface area $m^2\ g^{-1}$	Particle size, d_{50}, μm	Reversible capacity, $mAh\ g^{-1}$	First cycle efficiency, %
Osaka gas	MCMB 6-28	Soft Carbon, Graphitized at 2800 °C	4	5–7	295.3	92
	MCMB 10-28		1.96	12.13	326.9	92
	MCMB 25-28		1	25	334.0	92
	OMAC-15	Soft Carbon, Graphitized	2.59	15.31	364.9	96
	OMAC-21		N/A	N/A	356.0	93
Hydro-Quebec	SNG-12		8.59	12.10	370	86
Superior Graphite	LBG-1025	Natural Graphite	<4.9	19	360	91
Superior Graphite	SLA-1020	Thermally Purified Natural Graphite	<5.5	20	363	92
Superior Graphite	SLC-1015	Surface-Treated Natural Graphite	<1.2	25	352	92

1.4 Electrodes: Conductive Diluents and Binders

Typically, a lithium-ion cell electrode is a composite, also called a laminate, consisting of the following agglomerated components: (1) primary particles of active intercalation materials and (2) inactive materials coated and pressed (calendered) onto the metallic current collecting foil (Cu for the anode, Al for the cathode). The group of inactive materials typically consists of conductive additive(s) and polymeric binder.

The active material acts as the lithium-ion source and acceptor in the electrode, whereas the binder plays the role of binding the active materials and the conductive additive together and to the current collector. The most commonly used conductive diluents are carbon black and graphites with small particles (<6 μm) and large specific surface area (>50 $m^2\ g^{-1}$), which is more than 10–20 times higher than the specific area of the most common active electrode materials. The prevailing binder used in lithium-ion batteries is polyvinylidene difluoride (PVDF). Various grades of PVDF for lithium-ion batteries differ in their physical and chemical properties, such as molecular weight, and surface chemistry [39]. The high molecular weight of PVDF has reportedly improved the adhesion of electrode to current collector, as evidenced by the peel test results, as well as the capacity of the cell [40]. The surface chemistry of PVDF influences the interactions with the bound particles and affects binder's distribution homogeneity and level of coverage of active material particles by the polymer.

The conductive carbons do not contribute to the electrochemical process of the cell but is required to lower the internal cell resistance and therefore to optimize the power density and specific power of the cell. The binder is also inactive in cell reactions but ensures the cohesion of the electrode and the adhesion of the electrode components to the current collector. The careful optimization of electrode composition is crucial for high performance of lithium-ion cell in any application. The following composition-related trends have been identified:

(1) Increasing the content of conductive carbon contributes to high electronic conductivity of the electrode laminate. However, the excessive amount of a high surface area carbon additive might spoil the close contact between active material particles and bring the electronic conductivity of the electrode down.
(2) Excessive binder content PVDF in the electrode leads to a decrease in the ionic conductivity of the electrode due to the ion-blocking property of the ionic insulating binder. However, decreasing the binder content beyond a certain limit (specific to the materials used in the electrode) causes a decrease in the electronic conductivity and cycle life of the electrode due to the poor physical connection between the electrode particles and the current collecting foil.
(3) The influence of the electrode composition on the electrode performance can be related to the binder/conductive carbon ratio. At the optimized binder/conductive carbon ratio, volumetric energy density of the electrode can be improved by using less of all the inactive materials (binder and the conductive carbons) without loss of power density. However, a low total content of the inactive materials is not recommended because the electrode may have a mechanical integrity issue. The exact optimized binder/conductive carbon ratio is specific to the particular active and inactive materials. Nevertheless, the trend is universal for lithium-ion cathode containing oxide active materials and inactive carbonaceous diluents [41].
(4) As the specific energy and energy density of the cell is optimized by keeping the respective amount and volume fraction of conductive carbon as low as possible, there is bound to be a trade-off between cell energy and power.
(5) Besides the critical volume fraction, the surface area and morphology, compaction behavior, as well as the handling and processing of the conductive carbon are important aspects for the electrode engineering [42].
(6) For optimal results in terms of electrode resistivity, a homogeneous dispersion of carbon black in the electrode mass only can be achieved by adapting the mixing process to the specific properties of each carbon black, such as their surface chemistry. Carbon black with lower degree of graphitization and a higher concentration of hydrogen, oxygen, and carboxylic acid groups on the surface is easily dispersed, with dry mixing being sufficient for homogeneous distribution within the electrode. The lower specific surface area is the reason for the lower viscosities of solvent or aqueous dispersions of this carbon black. The lower amount of liquid required when preparing the coating dispersion during the electrode manufacturing and as a consequence, the shorter electrode drying times are process advantages expected for the application of this carbon black in lithium-ion battery electrodes [42].

(7) The role of carbon black in the electrode can be supported by adding the second, complementary type of conductive carbon: small-particulate graphite. Graphitic carbon additives allow for achieving higher compressed densities of the fabricated electrodes [43]. Higher densities of electrode decrease the distance between all of the electrode particulate components, both active and inactive and could also additionally improve electronic conductivity within the electrode by facilitating the electron tunneling or hopping effects. Additionally, the higher thermal conductivity of graphite is expected to improve heat dissipation in the electrode [43].

1.5 Electrolyte Solutions

The most common lithium-ion cell electrolytes are based on solutions of a lithium salt in the mixture of multiple organic solvents. Multiple solvents with very different physical and chemical properties are often used together to perform various functions simultaneously in a lithium-ion cell that is expected to perform in wide range of conditions (temperature, cycling rates, etc.). The most commonly used solvents are listed in Table 1.8 and salts are summarized in Table 1.9.

Usually, a single salt chemistry is used in a cell, because the performance advantages of multiple salts have not been clearly demonstrated to date and the suitable salt anion candidates are sparse. The ideal solvent should fulfill the following minimal criteria:

(1) It should dissolve salts to sufficient concentration, i.e., have a high dielectric constant.
(2) It should possess low viscosity, so that facile lithium-ion transport can occur.
(3) It should remain inert to all cell components, especially the charged surfaces of the cathode and the anode, during cell operation.
(4) It should remain liquid in a wide temperature range. In other words, its melting point should be low and its boiling point high.
(5) It should demonstrate a high flash point (safety), be nontoxic, and economical [44].

The best candidate for an electrolyte solute (lithium-ion salt) for rechargeable lithium-ion batteries should meet the following minimal requirements:

(1) It should be completely dissolved and dissociated in the solvent mixture.
(2) The solvated lithium cations should move within the media with high mobility.
(3) The anion should be inert to electrolyte solvents.
(4) The salt anion should be stable against oxidative decomposition at the cathode.
(5) Both the anion and the cation should remain inert toward all the cell components, such as separator, electrode substrates, and cell packaging materials.
(6) The anion should be nontoxic and stable against thermally induced reactions with electrolyte solvents and other cell components [44].

Table 1.8 The most common organic solvents in lithium-ion electrolytes

Solvent	Structure	MW, g mol^{-1}	T_m, °C	T_b, °C	T_f, °C	η, cP 25 °C	Dipole moment, debye	d, g cm^{-3}
Ethylene Carbonate, EC	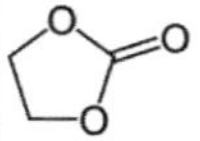	88	36.4	248	160	1.90 (40 °C)	4.61	1.321
Propylene Carbonate, PC	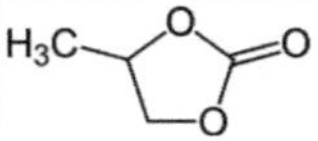	102	−48.8	242	132	2.53	4.81	1.200
Dimethyl Carbonate, DMC	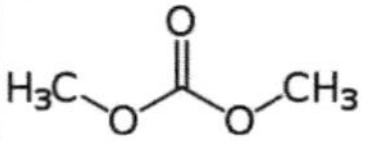	90	4.6	91	18	0.59 (40 °C)	0.76	1.063
Diethyl Carbonate, DEC	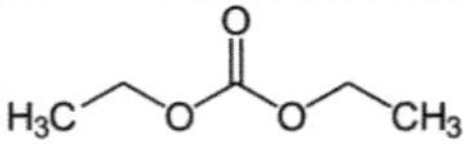	118	−74.3	126	31	0.75	0.96	0.969
Ethyl Methyl Carbonate, EMC	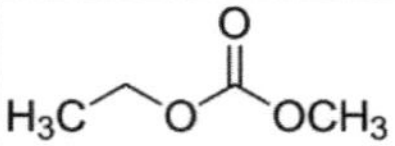	104	−53	110	23.9 [45]	0.65	0.89	1.006
Gamma-butyrolactone, γBL	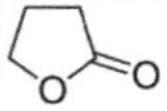	86	−43.5	204	97	1.73	4.23	1.199

Table 1.9 Common lithium salts (solutes) in lithium-ion electrolyte

Salt	Structure	MW, g·mol^{-1}	T_m, °C	$T_{decomp.}$, °C solid	σ, mS cm^{-1} in EC:DEC (3:7)
Lithium hexa fluorophos phate, $LiPF_6$		151.9	194 (d)	194 [46]	7.5 (1.0 M, 25 °C) [47]
Lithium bis(oxalato) borate, LiBOB	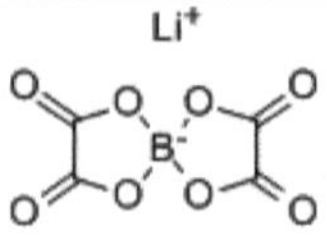	193.8	Decomposes before melting	293 [48]	4.2911 ± 0.0045 [49] (0.5 M, 25 °C)

Note Poor solubility of LiBOB in electrolyte (e.g., EC:DEC mixture) solvents does not allow for preparation of a 1.0 M solution

1.6 Porous Separators

Microporous polymer membranes are the most commonly used separators in lithium-ion batteries. Majority of the microporous polymer membrane separators are based on semicrystalline polyolefin materials, such as polyethylene (PE), polypropylene (PP), PE–PP blends and high-density polyethylene (HDPE)-ultrahigh molecular polyethylene (UHMWPE).

An emerging new group of lithium-ion separator comprises of inorganic composite membranes ("ceramic separators") with their excellent wettability and exceptional thermal stability properties. Most of ceramic separators are porous mats made of ultrafine inorganic particles bonded using a small amount of binder.

The essential function of a separator is to prevent the physical and electrical contact between cathode and anode while permitting Li^+-ions flow across the cell. The presence of a separator affects the battery performance, as well as energy and power densities, cycle life and safety but the separator material should not participate in any cell reactions (Table 1.1). The major requirements for separators used in rechargeable Li-ion batteries are listed in Table 1.10 [50].

1.7 Future Trends in Lithium-ion Cell Materials

The discussion of lithium-ion battery materials in this chapter focused on the well-established products. The research and development work on the cell components and materials is still very much active and many newly synthesized products are made available for testing and implementation. With time, market needs and the actual performance of these new materials in cells will separate the best candidates for future iterations of lithium-ion batteries. Currently, the most prevalent target requirements of batteries for portable electronics and transportation (electric vehicles, EV) applications are increased energy density and specific energy, closely

Table 1.10 Requirements for lithium-ion cell separators

Requirement	Notes	Testing method	Typical value
Chemical stability	Stability against the electrolyte and electrode materials: reductive and oxidative environments of a fully charged battery	Cycle life testing in full cell	Several thousands of cycles
Thickness	Lowest possible, too thin adversely impacts mechanical strength and safety	T411om-83 [51]	25.4 μm
Porosity	Necessary to hold liquid electrolyte for the ionic conductivity across the cell. Too high porosity adversely impacts the shutdown properties	ASTM [52] D-2873	40 %
Pore size	Uniform pore size and tortuous pore structure are desired	ASTM 1294	Submicron range
Permeability	Measured by taking time elapsed when a specific amount of air passes through a specific area of separator under specific pressure	ASTM D726	Varies
Mechanical strength	Tensile strength: • machine direction • transverse direction	ASTM: D-638	>1000 kg cm^{-2}
	Tear resistance	D-1004	
	Puncture strength	D-822	300 g
Wettability	Visual observation of a droplet of electrolyte on a separator	No standard test method	N/A
Dimensional stability	Flat when dry or soaked with electrolyte	No standard test method	N/A
Thermal shrinkage	Most polyolefin materials contain amorphous and crystalline phases with varying densities; at elevated temperature, thermal shrinkage occurs	Measurement of dimensional change upon heating, area based [53]	Maximum 5 % at 60 °C for 1 h
Shutdown properties	Observed by differential scanning calorimetry (DSC) or resistance change of the electrolyte-soaked membrane with temperature increase		130–165 °C
Cost	The cost of separator can constitute ~20 % of the total price of lithium-ion battery		

Table 1.11 Future or developmental materials for lithium-ion batteries

Material	Notes
Layered-layered composite cathode materials, $xLi_2MnO_3{\cdot}(1-x)LiMO_2$ (M = Mn, Ni, Co)	Promising high-capacity cathode material developed at the Argonne National Laboratory, capable of delivering discharge capacity in excess of 250 mAh g^{-1} [54]. Challenges are related to voltage fade/cycle life issues as well as development of the compatible, high-voltage electrolytes (~5 V)
Phosphate cathodes; $LiMPO_4$ (M = Mn, Ni, or Co), or Li3V2(PO4)3	Electronic conductivity remains a problem for all phosphates; Ni and Co-containing phosphates require electrolytes that are stable around 5 V
Silicon-based anode materials	Several manufacturers claim progress in pure Si and Si-based composite areas (mostly, Si-carbons). The challenges are mostly related to scale-up of Si-based anodes, development of compatible electrolytes and cycle life extension
Tin-based anode materials	The most commercially advanced anodes in this group are Sn–Co–C composites implemented by Sony Corporation in 2005 [55]. The developmental work is still ongoing to maximize the capacity and cycle life of this anode

followed by improved safety, low cost, and long cycle life. Most of the portable electronics and EV applications require medium power (~2C or lower discharge current for several seconds to minutes).

There are emerging opportunities in higher power applications, such as hybrid electric vehicles (HEV), but at the moment, Li-ion batteries are the disadvantaged HEV option compared to Ni/metal hydride batteries due to the cost and safety concerns.

Li-ion batteries for stationary applications in their present form are a questionable choice as there are competing technologies that have produced more viable options for these applications; for more information on competing technologies see Chap. 7 of this book. The (subjective) list of the most promising materials for future lithium-ion batteries is listed in Table 1.11, along with the perceived obstacles to full implementation of these technologies.

In general, for the next 5–10 years, the developments in lithium-ion battery materials will most likely continue being incremental. The development of silicon and tin-based composite anode materials is the "low hanging fruit" for the development activities, as the materials capable of 1,000–1,200 mAh g^{-1} are very feasible and the Sn-alloy and Si–C materials can be tailor-made to enter the current electrode manufacturing process with minimal adjustments. The next "pain point" of the lithium-ion technology is related to the electrolyte solutions that are systemically compatible with the novel anode and cathode materials and possess the extended voltage (above 5 V) range. The development of high voltage electrolytes would enable commercial implementation of an array of the cathode materials that are close to commercialization but halted due to the lack of suitable electrolyte, such as high voltage spinels (e.g., $LiMn_{1.5}Ni_{0.5}O_4$), thus further increasing the overall specific energy of lithium-ion batteries. The combinatorial chemistry and modeling efforts are expected to have the increasing impact on the development of the novel materials, allowing for the shortened time needed for the development and testing of multiple iterations of new materials (for electrodes and electrolytes alike). Finally, hybrid systems (Chap. 6) consisting of lithium-ion and other devices (such as supercapacitors, fuel cells, and other) are expected to proliferate in the near future.

References

1. Whittingham MS (2004) Lithium batteries and cathode materials. Chem Rev 104:4271–4301. doi:10.1021/cr020731c
2. Levasseur S, Menetrier M, Suard E, Delmas C (2000) Evidence for structural defects in non-stoichiometric HT-LiCoO2: electrochemical, electronic properties and Li-7 NMR studies. Solid State Ionics 128:11–24. doi:10.1016/S0167-2738(99)00335-5
3. Reimers JN, Dahn JR (1992) Electrochemical and insitu X-ray-diffraction studies of lithium intercalation in LixCoO2. J Electrochem Soc 139:2091–2097. doi:10.1149/1.2221184
4. Huggins RA (2009) Advanced batteries: materials science aspects. Springer, New York
5. Yamahira T, Kato H, Anzai M (1991) US Patent 5,053,297
6. Goodenough JB, Mizushima K (1981) US Patent 4,302,518, (1982) US Patent 4,357,215
7. Mizushima K, Jones PC, Wiseman PJ, Goodenough JB (1980) LixCoO2 $(0 < x \leq 1)$: A new cathode material for batteries of high energy density. Mater Res Bull 15:783–789. doi:10.1016/0025-5408(80)90012-4
8. Lu CH, Yeh PY (2002) Microstructural development and electrochemical characteristics of lithium cobalt oxide powders prepared by the water-in-oil emulsion process. J Eur Ceram Soc 22:673–679. doi:10.1016/S0955-2219(01)00366-1
9. MacNeil DD, Dahn JR (2001) The reaction of charged cathodes with nonaqueous solvents and electrolytes - I. Li0.5CoO2. J Electrochem Soc 148:A1205–A1210. doi:10.1149/1.1407245
10. Nishi Y (2001) Lithium ion secondary batteries; past 10 years and the future. J Power Sources 100:101–106. doi:10.1016/S0378-7753(01)00887-4
11. Chen Z, Ellenwood R (2010) US Patent Application US20100176352
12. B. Scrosati (2000) Recent advances in lithium ion battery materials. Electrochim Acta 45:2461–2466. doi:10.1016/S0013-4686(00)00333-9
13. http://www.iucr.org/education/pamphlets. Accessed 1 March 2013
14. Pecharsky VK, Zavalij PY (2005) Fundamentals of powder diffraction and structural characterization of materials. Springer, New York
15. Cho J, Kim YJ, Kim JT, Park B (2001) Zero-strain intercalation cathode for rechargeable Li-ion cell. Angew Chem Int Ed 40:3367-3369. doi:10.1002/1521-3773(20010917)40:18<3367::AID-ANIE3367>3.0.CO;2-A
16. Wang ZX et al (2002) Electrochemical evaluation and structural characterization of commercial LiCoO2 surfaces modified with MgO for lithium-ion batteries. J Electrochem Soc 149:A466–A471. doi:10.1149/1.1456919
17. Kim Y, Kim D, Kang S (2011) Experimental and first-principles thermodynamic study of the formation and effects of vacancies in layered lithium nickel cobalt oxides. Chem Mater 23:5388–5397. doi:10.1021/cm202415x
18. Kim JM, Chung HT (2004) The first cycle characteristics of Li[Ni1/3Co1/3Mn1/3]O2 charged up to 4.7 V. Electrochim Acta 49:937–944. doi:10.1016/j.electacta.2003.10.005
19. Li DC et al (2004) Effect of synthesis method on the electrochemical performance of LiNi1/3Mn1/3Co1/3O2. J Power Sources 132:150–155. doi:10.1016/j.jpowsour.2004.01.016

20. Chen YH, Wang CW, Zhang X, Sastry AM (2010) Porous cathode optimization for lithium cells: Ionic and electronic conductivity, capacity, and selection of materials. J Power Sources 195:2851–2862. doi:10.1016/j.jpowsour.2009.11.044
21. Dahbi M, Saadoune I, Gustafsson T, Edström K (2011) Effect of manganese on the structural and thermal stability of Li0.3Ni0.7-yCo0.3-yMn2yO2 electrode materials (y = 0 and 0.05). Solid State Ionics 203:37–41. doi:10.1016/j.ssi.2011.09.022
22. Feng J et al (2012) Electrochemical property of LiMn2O4 in overdischarged conditions. Funct Mater Lett 5:1250028. doi:10.1142/S1793604712500282
23. Padhi AK, Nanjundaswamy KS, Goodenough JS (1997) Phospho-olivines as positive-electrode materials for rechargeable lithium batteries. J Electrochem Soc, 144:1188–1194. doi:10.1149/1.1837571
24. Reddy TB, Linden D, eds (2010) Linden's handbook of batteries 4th edition. McGraw-Hill, New York
25. Li G, Yang Z, Yang W (2008) Effect of FePO4 coating on electrochemical and safety performance of LiCoO2 as cathode material for Li-ion batteries. J Power Sources 183:741–748. doi:10.1016/j.jpowsour.2008.05.047
26. Xu F et al (2012) Failure Investigation of LiFePO4 cells under overcharge conditions. J Electrochem Soc 159:A678–A687. doi:10.1149/2.024206jes
27. Yamada A, Chung SC, Hinokuma K (2001) Optimized LiFePO4 for lithium battery cathodes. J Electrochem Soc 148:A224–A229. doi:10.1149/1.1348257
28. Ravet N et al Abs. 127, The Electrochemical Society and the Electrochemical Society of Japan Meeting Abstracts, vol. 99–2, Honolulu, HI, 17–22 October 1999
29. Tarascon JM et al (2010) Hunting for better Li-based electrode materials via low temperature inorganic synthesis. Chem Mater 22:724–739. doi:10.1021/cm9030478
30. Croce F et al (2002) A novel concept for the synthesis of an improved LiFePO4 lithium battery cathode. Electrochem Solid-State Lett 5:A47–A50. doi:10.1149/1.1449302
31. Rho YH, Nazar LF, Perry L, Ryan D (2007) Surface chemistry of LiFePO4 studied by Mossbauer and X-ray photoelectron spectroscopy and its effect on electrochemical properties. J Electrochem Soc 154:A283–A289. doi:10.1149/1.2433539
32. Chung SY, Bloking JT, Chiang YM (2002) Electronically conductive phospho-olivines as lithium storage electrodes. Nat Mater 1:123–128. doi:10.1038/nmat732
33. Herle PS, Ellis B, Coombs N, Nazar LF (2004) Nano-network electronic conduction in iron and nickel olivine phosphates. Nat Mater 3:147–152. doi:10.1038/nmat1063
34. Beguin F, Frackowiak E, eds (2009) Carbons for Electrochemical Energy Storage and Conversion Systems 1st edition. CRC Press, Boca Raton
35. Ridgeway P et al (2008) Effect of vinylene carbonate on graphite anode cycling efficiency. Electrochem Soc Trans 13:1–7
36. Zaghib K et al, Low-cost graphite and olivine-based materials for Li-ion batteries. BATT Review Meeting. Washington, DC February 25–28 2008
37. Barsukov IV et al, eds (2006) New Carbon Based Materials for Electrochemical Energy Storage Systems. Springer NATO Science Series. Dordrecht, Netherlands
38. Hirota N, Itabashi T, Maki S (2011) US Patent Application, 20110159360
39. Yoo M, Frank CW, Mori S, Yamaguchi S (2004) Interaction of poly(vinylidene fluoride) with graphite particles. 2. Effect of solvent evaporation kinetics and chemical properties of PVDF on the surface morphology of a composite film and its relation to electrochemical performance. Chem Mater 16:1945–1953. doi:10.1021/cm0304593
40. Amin-Sanayei R, Heinze R, NASA Aerospace Battery Workshop Presentation. Huntsville, AL 2009
41. Zheng H et al (2012) Cooperation between active material, polymeric binder and conductive carbon additive in lithium ion battery cathode. J Phys Chem C 116:4875–4882. doi:10.1021/jp208428w
42. Zheng H et al (2010) Cathode performance as a function of inactive material and void fractions. J Electrochem Soc 157:A1060–A1066. doi:10.1149/1.3459878

43. Spahr ME et al (2011) Development of carbon conductive additives for advanced lithium ion batteries. J Power Sources 196:3404–3413. doi:10.1016/j.jpowsour.2010.07.002
44. Xu K (2004) Nonaqueous liquid electrolytes for lithium-based rechargeable batteries. Chem Rev 104:4303–4417. doi:10.1021/cr030203g
45. Material Safety Data Sheet (MSDS) for EMC (CAS Number 623-53-0, 99 %), Sigma-Aldrich
46. Zinigrad E et al Abs. 245, The 210th Electrochemical Society Meeting Abstracts, Cancun, Mexico, October 29-November 3 2006
47. Lex-Balducci A et al (2010) Lithium borates for lithium-ion battery electrolytes. Electrochem Soc Trans 25:13–17
48. Zinigrad E et al (2007) On the thermal behavior of Li bis(oxalato)borate LiBOB. Thermochim Acta 457:64–69. doi:10.1016/j.tca.2007.03.001
49. Smart MC et al (2002) Performance characteristics of lithium ion cells at low temperatures. IEEE Trans Aerosp Electron Syst Mag 17:16-20. doi:10.1109/MAES.2002.1145732
50. Zhang SS (2007) A review on the separators of liquid electrolyte Li-ion batteries. J Power Sources 164:351–364. doi:10.1016/j.jpowsour.2006.10.065
51. Test method developed under auspices of the Technical Association of the Pulp and Paper Industry
52. American Society for Testing and Materials (ASTM)
53. Jeong HS, Hong SC, Lee SY (2010) Effect of microporous structure on thermal shrinkage and electrochemical performance of Al2O3/poly(vinylidene fluoride-hexafluoropropylene) composite separators for lithium-ion batteries. J Membr Sci 364:177–182. doi:10.1016/j.memsci.2010.08.012
54. Johnson CS et al (2008) Synthesis, characterization and electrochemistry of lithium battery electrodes: xLi2MnO3•(1-x)LiMn0.333Ni0.333Co0.333O2 ($0 \leq x \leq 0.7$). Chem Mater 20:6095–6106. doi: 10.1021/cm801245r
55. Sony news release. http://www.sony.net/SonyInfo/News/Press/200502/05-006E/index.html. Accessed 29 April 2013

Chapter 2
Predicting Materials' Performance

Boris Ravdel

Abstract One of the most vital challenges in lithium-ion battery engineering and manufacturing is related to the accurate prediction of cell's performance in various conditions (operating and storage temperature range, cycling rates, prolonged cycling, etc.). Macro-kinetic modeling of the processes occurring in the cell made tremendous progress during recent years. At the same time, substantial lack of data on intrinsic kinetic parameters, such as diffusion coefficients, exchange current, and their dependence upon lithium content in the solid materials restrains implementation of modeling. A practical approach to determining the proper form of the kinetic equations is presented in this chapter. This approach is based on interpretation of electrochemical measurements with respect to the structure of the electrode materials, and establishing the kinetic parameters, taking into account the connection between thermodynamic functions and kinetic parameters. The methodology refuses employing the concept of thermodynamic activity, but intensively uses empirical data of open circuit potentials (OCP) of the battery electrodes at varying degree of charge (DoC) measured at various temperatures. These data can also be useful for better understanding and prediction of most favorable voltage range for prolonged battery cycling life, and decrease the risk of catastrophic failures related to overtaxing the battery's active materials. Several examples of practical accomplishments of this approach are included in this chapter.

2.1 Editor's Note

For the reader, whose everyday "use" of electrochemistry is limited to the rather utilitarian consideration of simple formulas and approximated processes ongoing in the electrochemical cell (myself included), the following chapter may at first

B. Ravdel (✉)
Yardney Technical Products, Inc., 2000 South County Trail, East Greenwich, RI 02818-1530, USA
e-mail: bravdel@yardney.com

M.K. Gulbinska (ed.), *Lithium-ion Battery Materials and Engineering*, Green Energy and Technology, DOI 10.1007/978-1-4471-6548-4_2

seem challenging. However, the included information is well worth the effort of brushing up on the principles and applications of fundamental electrochemistry.

Simply, the derivations presented here and in the related cited literature as well as the proposed experimental approach have direct use in speciality lithium-ion cells. The described analyses and their results have already assisted in fielded products with tens of thousands of delivered cycles (Fig. 2.1).

Another possible use of this material for the interested reader relates to the topic of lithium-ion cell modeling, which relies on similar foundations to the equations presented below, which are explicit despite being simplified. The simplification was done to convey the general concept; further reading is certainly encouraged.

2.2 Introduction: Macro and Intrinsic Kinetics

Generally, the electrochemical reaction is a heterogeneous, multi-step process. These steps can be consecutive or parallel; they can include homogeneous chemical reactions, transport processes, adsorption phenomena, crystals' nucleation and growing, as well as formation of new phases, etc. However, one essential step, always required to occur in the electrochemical reaction is the electron transfer through the electrolyte solution-electrode phase boundary. Thus, the electronic conductivity of at least one phase is crucial for the reaction to proceed. The overall reaction can involve several electrons; the electrons being transferred simultaneously or stepwise. In the latter case, other steps sometimes take place between the electron transfer steps.

Text Box 2.1 Heterogeneous reactions [1]

Heterogeneous reactions constitute a class of chemical reactions where the reactants are components of two or more phases (solid and gas, solid and liquid, two immiscible liquids), or in which one or more reactants undergo chemical change at an interface, e.g., on the surface of a solid electrode.

An example of the simplest (in the sense of the number of kinetic parameters) electrochemical reaction is reduction of silver ions (Ag^{+}) from a dilute aqueous solution of a well soluble silver salt (e.g., nitrate) in the presence of excess of an indifferent salt (e.g., potassium nitrate) on a liquid silver-mercury alloy (also called amalgam) electrode. Besides the transfer of a single electron, only diffusion steps are involved in this process. The entire reaction can be very well modeled and the kinetic parameters are determined experimentally with high level of accuracy. The information gleaned while analyzing the mechanism of silver ion reduction can be used in elucidating more complex, multi-step, multiphase processes, such as the electrochemical reaction in a lithium-ion cell.

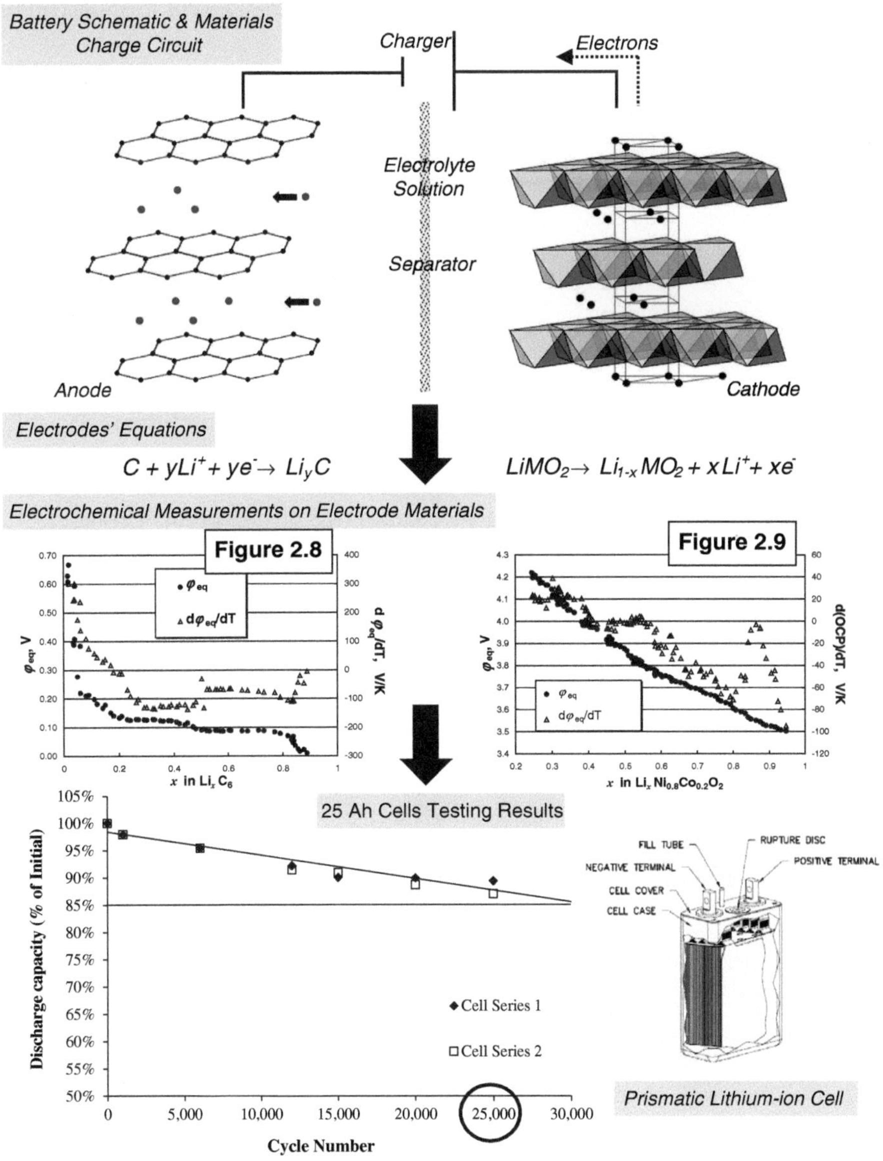

Fig. 2.1 From fundamental properties of electrode materials to cell performance optimized for the fielded product

Active materials of lithium-ion electrodes[1] can be represented by the formula Li_xZ ($0 < x < 1$). Here, x corresponds to that part of lithium content, which can reversibly leave the material (mobile lithium, in the form of Li^+ ions). The moiety

[1] That is, the active materials of the electrodes employed in a lithium-ion cell.

Z may contain lithium and the stoichiometric indices in the explicit formula can be fractional (Text Box 2.2).

Text Box 2.2 Comments on selected moieties found within lithium-ion cathodes

The term *moiety* in this text refers to the functional group of atoms found within the formula of electrode material. An example of the moiety Z (as in Li_xZ) is the ($Li_{0.45}CoO_2$) group in the $Li_{0.55}(Li_{0.45}CoO_2)$ notation of $LiCoO_2$ formula. The notation gives information on the content of mobile lithium in the compound.

Many partially lithiated (i.e., when $x < 1$) active materials in lithium-ion cells can be considered as solid solutions of mobile lithium (guest) in the host material—Z. Such host–guest material can be described by the similar set of thermodynamic equations as these used to characterize an amalgam. However, the mechanism of electrochemical reactions proceeding in a real cell is much more complex than in the case of the reaction of silver amalgam formation. There are many factors complicating the overall process:

- porous structure of the electrodes,
- low electron conductivity (as a rule) of the active material,
- presence of alien materials such as polymer binder and conductive diluents.

As the result, the basic electrochemical relationships binding the rate of the electrochemical reaction with the electrode potential are insufficient for the complete description of the cell processes. Instead, rather sophisticated macro-kinetic models have been developed. On the other hand, all models require quantitative data on several kinetic parameters. In the simplest case, it is enough to know the rate constant of the charge transfer step, or exchange current (EC), and diffusion coefficients (DC) in solid phase and electrolyte solution, assuming that both EC and DC depend on concentration.

Another term frequently used in modeling of electrochemical processes is polarization. Polarization, i.e., the shift of electrode potential from equilibrium state, is commonly employed as the driving force of electrochemical reaction. Consequently, the quantitative description of the discharging and charging processes in lithium-ion cell requires the knowledge of the equilibrium potential (EP) of electrodes as functions of degree of charge and temperature. In conventional electrochemistry, Nernst equation is used to determine the EP at given concentration and temperature. However, no general type of the dependence of EP on concentration and temperature is known or even suggested for lithium-ion electrode materials. Therefore, for all new cell chemistries (e.g., the active electrode materials described in Chap. 1) these relationships must be determined experimentally.

The overreaching aim of the work described here was to establish a generalized description of the processes in lithium-ion cell that would apply regardless of the exact chemistry of cell materials or nonequilibrium conditions. The approach taken in this publication relied on experimental observations, and the ultimate goal was the practical application. The application of this chapter would be the prediction of the complex electrochemical behavior of materials in an actual lithium-ion cell, based on the specific, easy to measure set of parameters.

In the course of this chapter, the simplest path of the overall process (diffusion, charge transfer, diffusion) is discussed, established by using phenomenological approach and focusing on the practical applications. Some conceptual challenges persist even at this simplified level, and are enumerated in the following text. The developed approach is applicable not only to the lithium-ion electrodes, but also to all electrochemically active solid phases (EASP).[2] In the following sections, the term EASP refers to a solid substance capable of electrochemical reduction or oxidation by species from another phase. Usually, the other phase is a liquid solution, but the participation of a second solid phase is possible. Oxidation or reduction processes involve changing the composition, as well as chemical and physical properties of EASP due to passing electricity through the EASP/solution boundary. It is important to note that the electrochemical and some accompanying chemical reactions in solid phase proceed at nonsteady-state conditions.

The interest in the analysis of the dependencies of equilibrium potential on composition of cathode materials for lithium-metal cells appeared in the late-1970s [2–8] where phase composition and phase transitions of oxides and halcogenides of transient metals upon lithiation were discussed. The usefulness of the simultaneous scrutiny of the equilibrium potential together with its temperature coefficient was first proved in several works [9–13] published soon after. The approach to the calculation of kinetic parameters using the thermodynamic data, which is the subject of this chapter, has been proposed [14–16] later. In early 2000, new interest in the method has arisen, both in the thermodynamics of the processes within the electrodes for lithium-ion cells [17–22] and in the connection between thermodynamic functions and kinetic parameters [23]. In the series of recent works, M. Bazant [24] described the development of the fundamental theory of electrochemical kinetics and charge transfer applied to lithium iron phosphate (LFP).

2.3 Thermodynamics of Solid Electrodes

The knowledge of EP and the EP's dependence on electrode active material's composition at various temperatures is essential for calculations in the field of applied electrochemistry. Moreover, the EP (φ) is directly proportional to the Gibbs energy (ΔG) for the electrochemical reaction, and EP's temperature

[2] The application of concepts discussed in this chapter can be extended onto alloy electrodes where insertion processes can occur, to electrochemical alloys' formation, etc.

coefficient (TCEP, dφ/dT) is proportional to entropy (ΔS) of the process. In other words, the EP and TCEP are Gibbs energy and entropy expressed in electrical units. Parallel scrutiny of both thermodynamic functions is a powerful tool of physicochemical analysis.

The concepts of EP and related open circuit potential and voltage (OCP and OCV), and electromotive force (emf) are well described in many textbooks on physical chemistry or electrochemistry. However, proving that the measured OCV (OCP) is indeed EP requires an approach that is specific to the particular electrochemical process being analyzed. For the EASP electrodes, a good confirmation of the OCP $\equiv$ EP relationship is the stability of the OCV versus lithium-metal (Li | Li^+) reference electrode over time and independence upon lithium ion's concentration. According to the IUPAC [25] rule, the electrochemical circuit for such measurements has the form

$$\mathrm{Li}_x\mathrm{Z} \,\middle|\, \mathrm{Li}^+(\mathrm{solvent}) \,\middle|\, \mathrm{Li} \tag{2.1}$$

meaning that Li | Li^+ is the reference electrode, at the potential of 0 V. As with other reference electrodes, when citing the OCV values measured versus Li | Li^+, the type of the reference electrode used in measurements should be stated because of the reference's impact on the measurements.

2.3.1 Ideal and Real Solutions

Bazant's theoretical work [24] exploits extensive knowledge of properties of LFP nanoparticles. LFP is a well-known and researched cathode material for lithium-ion batteries. The known characteristics of LFP include specifics of phase separation conditions, lithium-ion transport mechanism, elastic coherency strain, anisotropic nucleation and growth, interfacial energies, etc. Bazant's work and other similar approaches describing electrochemical processes in nonequilibrium conditions are crucial to the progress of applied electrochemistry in general and specifically, to forwarding developments in the field of intercalation materials-based batteries. The following section of this chapter compares the ideal and real systems, and the associated electrochemical solutions/equations.

If a reversible redox reaction:

$$\mathrm{Ox} + ne \leftrightarrows \mathrm{Red} \tag{2.2}$$

proceeds on an electrode in the so-called "ideal solution", the dependence of the EP upon reactants' concentrations, [Ox] and [Red], follows the Nernst equation:

$$\varphi = \varphi^\circ + \frac{RT}{nF}\ln\frac{[\mathrm{Ox}]}{[\mathrm{Red}]} \tag{2.3}$$

where φ° is the standard potential at standard values of the both reactants' concentrations, and R, T, and F denote the universal gas constant, the absolute temperature, and the Faraday constant, respectively. In fact, Eq. 2.3 is a transition of the formula for the chemical potential of ideal gas:

$$\mu = \mu^{\circ} + RT\ln P \tag{2.4}$$

μ° is [26] the standard value of the chemical potential at standard pressure p°, and the dimensionless value $P = p/p^{o}$ is the ratio of the actual gas pressure over the standard pressure p°. The real dependence of the EP upon concentration follows the Nernst relationship only at low concentrations (below 0.001 mol L^{-1} in aqueous solutions).

Similarly, Eq. 2.4 poorly describes the behavior of gases at pressure above 50 atmospheres. In order to retain the appearance of the ideal-gas thermodynamic equations, but render them applicable to real gases, an American chemist G.N. Lewis introduced the concept of fugacity in 1901 [27] The ideal gas pressure p and fugacity f is related through the dimensionless fugacity coefficient $g = f/p$. The fugacity coefficient has clear physical meaning, and describes the deviation in behavior of a real gas compared to an ideal gas.

Indeed, all gases follow the general equation

$$\left(\frac{\partial \mu}{\partial p}\right)_T = \underline{v} \tag{2.5}$$

where $\underline{v}$ is the gas molar volume. For an *ideal* gas, the equation of state has the form:

$$p\underline{v} = RT \tag{2.6}$$

For a *real* gas, other equations of state must be used, e.g., van der Waals, Berthelot, Diterichi, or Kammerling-Onnes [28]. Integration of Eq. 2.5 using Eq. 2.6 leads to the relationship shown in Eq. 2.4. In this case, standard chemical potential appears formally as the integration constant. Of course, integrating of Eq. 2.5 with a real gas equation of state will produce other forms of chemical potential equation than in the case of the ideal gas equation; the difference between the outcomes of integration uniquely defines the fugacity coefficient.

There is no single equation of state for the electrolyte solutions, but similarly to the concept of fugacity, the concept of thermodynamic activity is commonly engaged. The activity exhibits dependence on concentration.

By the IUPAC [25] definition, the activity (a) is the anti-logarithm of the chemical potential and does not constitute a new physical entity:

$$a = \exp\left(\frac{\mu - \mu^{\circ}}{RT}\right) \tag{2.7}$$

The concept of activity (Eq. 2.7) merely puts into operation *another function of chemical potential* and converts the chemical potential into a form that is convenient for calculations. Specifically, the thermodynamic functions can be calculated using the formulae derived for ideal gases by substitution of the pressure with the activity. In other words, the form of the definition (Eq. 2.7) is dictated by the wish to keep the formal similarity between the expressions for chemical potentials of ideal gas and a solution component. The activity is reciprocally bound with the standard value of the chemical potential. Since the standard chemical potential is equal to the chemical potential of the species in the *hypothetic* solution in which both activity and activity coefficient are equal to 1, the direct measurement is therefore impossible. Both standard potentials and activities (activity coefficients) must be calculated using one of several available modeling approaches. Computational procedures are not simple, and some problems are essentially irresolvable (e.g., the problem of the activity coefficient of a single ion; instead the *mean ion activity coefficient* is used). Nevertheless, the values of activity coefficients have been determined for many ions in various aqueous solutions. However, the collection of activity coefficients for ions in nonaqueous solutions is much scarcer.

The above-mentioned considerations pertained to computational challenges with analyzing electrochemical processes involving real gases and solutions. New set of difficulties arises with attempts of determining the activities (or activity coefficients) for solid phases comprised of multiple components. For example, the *empirical* Raoult and Henry laws that bind the concentrations of the solutions' components to their vapor pressures can hardly be practically applied for solids. Similarly, van't Hoff's law of osmotic pressure that formally implies the equation of state of ideal gas to the solution [28] is not applicable to solids. Indeed, the pressure of the solid solution (e.g., metal alloy) components is usually immeasurable within wide temperature range, and osmotic pressure is not detected in solids. This aspect is very prominent in EASP where lithium is dispersed in the host matrix.

Last but not the least problem with unified approach to electrochemical processes across various states of matter is the use of molar fraction [25] as the unit of concentration in electrochemical equations. Such choice of the concentration unit implies that the EP would exhibit *a very simple dependence* on the presence of the so-called indifferent electrolyte. By this logic, the potential of $Ag \mid Ag^+$ electrode in aqueous solution (e.g., 0.1 mol L^{-1} of silver nitrate) should depend on the addition of sodium nitrate since the molar fraction of silver ions is changed by the addition of the indifferent salt (KNO_3) even if the molar-volumetric concentration of Ag^+ ions remains constant. Moreover, the substitution of sodium salt with any other apparently indifferent salt (potassium, ammonium, alkyl-ammonium salt, etc.) is expected to shift the EP to new values. However, the indifferent (or supporting) electrolyte in common electrochemical practice is considered to be only affecting (stabilizing) the activity coefficient. On the other hand, the unanswered question persists whether the potentials of *ideal* silver-mercury and silver-gold alloy electrodes in silver nitrate solution are *equal* when silver *mole fractions* are *equal*, or when the silver *molar-volumetric concentrations* are *equal*.

The above-mentioned issues also pertain to the selection of the standard state of EASP. As a matter of fact, on $Li_xZ \mid Li^+$ electrode, either at $x = 0$ or at $x = 1$, no equilibrium conditions corresponding to lithium exchange can be established. The reasoning presented earlier illustrates that the application of the concept of thermodynamic activity to the EASP is impractical and may not be attainable.

2.3.2 Equilibrium Potentials on EASP

In order to define an alternative approach to employment of thermodynamic activity, a few basic assumptions must be established. First, let the oxidation or reduction reaction of EASP proceed with participation of a univalent cation from the electrolyte (e.g., Li^+). In reaction:

$$Z + xLi^+ + xe \leftrightarrows Li_xZ \tag{2.8}$$

Li_xZ and Z are reduced and oxidized forms of EASP, while $x \leq 1$. The symbol "$\leftrightarrows$" indicates that the reaction can proceed in both directions. However, in order to achieve the potential-determining reaction, it is more convenient to rewrite Eq. 2.8 so that the Li^+ coefficient equals 1. The modified equation is

$$pZ + Li^+ + e^- \leftrightarrows LiZ_p \tag{2.9}$$

Obviously, the EP can establish only on partially discharged electrode, that is, at $1 < p < 0$ (or $0 < x < 1$), but the fact that steady state OCV (OCP) is equal to the true EP remains yet to be proven.

Modifying the proposed EP-determining reaction on the $LiZ_p \mid Li^+$ electrode so that Eq. 2.9 is achieved clarifies the fact that the character of dependence of the electrode's EP *upon lithium ions' concentration in the solution* is the same as that of Li(metal)|Li^+ electrode (Eq. 2.9).[3]

$$\frac{\partial \phi_{\mathrm{EASP}}}{\partial [Li^+]} = \frac{\partial \phi_{Li(\mathrm{metal})}}{\partial [Li^+]} \tag{2.10}$$

Moreover, the equality of EP dependence on Li^+ concentration in solution with the Li | Li^+ metal electrode relationship is regardless of the application of the Nernst's law and the units in which lithium ion concentration in the solution is expressed. Thus, if the experimental measurements confirm the independence of the cell's OCV (Eq. 2.11) upon lithium ion concentration in the solution, the potential on the EASP can be treated as EP.

$$Li \left| Li^+(\mathrm{solvent}) \right| Li_xZ \tag{2.11}$$

2.3.3 The Dependence of Equilibrium Potential on DoC of EASP

There are several possible types of products forming during reduction of EASP. These various products differ in the nature of dependence of their EP upon DoC of EASP. For convenience and simplicity, the applicability of the Nernst equation to

[3] Lithium electrode $Li^+ + e \leftrightarrows Li$ in the discussed aprotic solutions is assumed to be at equilibrium state.

the systems described below is assumed, and the use of standard electrolyte solution is implied. Such approach does not affect the logic of exposition, but makes the description more illustrative. The four different EASP reduction products and the associated EP-DoC relationships are discussed below:

i. Lithium is inserted into the host matrix forming solid solution of *lithium atoms* within certain composition range. An example of this product is lithium insertion into aluminum matrix; solid solution is forming up to 9×10^{-4} at % of lithium [29]). Aluminum does not participate in the electrochemical reaction but acts as a solvent in the second phase, i.e., $x\mathrm{Li}^+ + \mathrm{Al} + xe^- \leftrightarrows x\mathrm{Li}\{\mathrm{Al}\}$.[4] The product can be expressed by the gross formula $\mathrm{Li}_x\mathrm{Al}$, which corresponds to the solution of Li^0 in aluminum with the resulting molar fraction of $y = x/(1 + x)$. The correlation between molar fraction and molar-volumetric concentration is nonlinear due to the dependence of the product density upon composition.
The potential-determining reaction is

$$\mathrm{Li}^+ + \mathrm{e}^- \leftrightarrows \mathrm{Li} \tag{2.12}$$

and the corresponding Nernst equation takes the form

$$\varphi = \varphi^o_{Li^+/Li^0\{Al\}} + \frac{RT}{zF}\ln\frac{c^\circ}{c} = \varphi' - \frac{RT}{F}\ln c \tag{2.13}$$

where c and c° are concentration and standard concentration of lithium in aluminum, respectively, whereas $z = 1$.
The EP from (2.13) can be represented as the explicit function of the DoC utilizing the relationship $y = cM_2/(cM_2 + \rho - cM_1) = x/(1 + x)$, where M_1 and M_2 are molar masses of the solute (lithium) and solvent (aluminum), and where ρ is the solution density (a known function upon composition). Though $d\varphi/d \ln x \neq RT/nF$, the corrective factor can be easily calculated.

ii. Lithium is inserted into the host matrix, forming solid solution and becoming a *part of a new, non-stoichiometric compound* within certain composition range. An example of this product is the formation of lithium niobium diselenide—$\mathrm{Li}_x\mathrm{NbSe}_2$ where solid solution is forming within the entire composition range ($0 < x < 1$ [30, 31]): $\mathrm{NbSe}_2 + x\mathrm{Li}^+ + xe \leftrightarrows \mathrm{Li}_x\mathrm{NbSe}_2$
The potential-determining reaction is

$$p\mathrm{NbSe}_2 + \mathrm{Li}^+ + \mathrm{e}^- \leftrightarrows \mathrm{Li}(\mathrm{NbSe}_2)_p \tag{2.14}$$

In this case, the EP depends upon DoC according to formula:

$$\varphi = \varphi' - \frac{RT}{F}\ln\frac{1-x}{x} \tag{2.15}$$

---

[4] Such case is not fully consistent with the previously given definition of EASP. This inclusion is justified by simplicity and completeness of the system description.

and

$$\frac{d\varphi}{d\ln x} = \frac{RT}{F}x(1-x)$$

That is, cases i and ii are *distinct using thermodynamic experimentation.*

iii. Lithium is inserted into the host matrix, forming solid solution *of non-stoichiometric, lithium-containing compound dissolved in the host matrix.* Such result was proposed in the work of Pozin et al. [10] for the product of reduction of manganese (IV) oxide within the range $0 < x < 1$: $MnO_2 + xLi^+ + xe^- \leftrightarrows Li_xMnO_2$
The potential-determining reaction is

$$p(x)MnO_2 + Li^+ + e \leftrightarrows (LiMnO_2)_{p(x)} \quad (2.16)$$

however, in reality, this process is best characterized by multiple equilibriums; different equilibrium achieved for each value of x. Accordingly, each point is characterized by a separate Nernst's equation:

$$\varphi = \varphi'(x) - \frac{RT}{F}\ln\frac{1-x}{x} \quad (2.17)$$

It is essential that the standard potential depends upon DoC: $\varphi^\circ = \varphi^\circ(x)$. Thus, the shift of EP is caused by the second (quantitative) component of the Eq. (2.17), and by the change of the standard potential due to the change of the nature of the product.
Similar phenomena are observed in liquid solutions where stepwise complex formation occurs. This solution can be characterized by the mean coordination number $\bar{n}$ [32]. When the ligand-metal ratio changes (e.g., initially, solution contains only ligand in certain concentration [L]; then, by titration, metal concentration [M] is increasing). In this case, both the total complex's concentration and mean coordination number undergo changes as the complex formation reaction proceeds. This process can be described as the change of the complex $ML_{\bar{n}}$ concentration with simultaneous altering of its composition. The shift of the potential $\Delta\varphi$ of metal (M) electrode versus the potential in the complex-free solution of metal ions is illustrated by the relationship

$$\Delta\varphi = \varphi^\circ(\bar{n}) - \frac{RT}{zF}\ln[ML_{\bar{n}}] \quad (2.18)$$

The physical meaning of this formula can be explained if the solvation of metal ions in highly concentrated electrolyte solutions (where concentration of metal ions is comparable to the reciprocal molar volume of the solvent) is considered instead of formation of complexes.

iv. Lithium insertion leads to formation of a set of new solid products where *one or more product(s) is/are lithium-containing compound(s).* An example of this set of products is the result of the reduction of copper(II) oxide [33]. In this case, the overall reaction is $CuO + 2Li^+ + 2e^- \rightarrow Cu + Li_2O$ or, if copper(I) oxide is taken into account as intermediate product, the resulting reaction is:

$2CuO + 2Li^+ + 2e^- \rightarrow Cu_2O + Li_2O$. Note that *one-directional arrows* connect the sides of the equations meaning that the reaction is irreversible. The formation of the set of products with various degree of lithiation is the most difficult case for electrochemical examination, due to several factors, listed subsequently. Let copper oxide transform according to the following scheme: $CuO \rightarrow Cu_2O \rightarrow Cu$ with accompanying formation of Li_2O. While copper(I) oxide is forming, the chemical potential remains constant; dramatically changing only when copper(II) oxide is exhausted and metal copper begins appearing. For that reason, two plateaus of EP are expected on the potential versus time plot, with virtually instant transition from one to another. However, it is not clear if the EP can be established on the electrode comprised of stoichiometric compounds, such as CuO, Cu_2O, $Li_2O \mid Li^+$.

The second difficulty is the verification of the presence of the true dependence of EP upon the DoC, since the dependence may be merely weak and indistinguishable from the experimental noise. It will be shown in the next section of this chapter that the measurement of temperature coefficient of the EP can shed some light on this problem.

The last difficulty is inherently related to the feasibility of nonequilibrium electrochemical measurements, which will be discussed later. Even very small disturbance of the electrode equilibrated within the plateau (such as applying a few-mV potential step or harmonic wave, or low current pulse) dramatically changes the electrode's surface composition, easily bringing the composition outside of allowed limits.

However, if the equation of state is unknown (i.e., the Nernst equation is inadequate), it is difficult to determine by the shape of the EP(x) curve, which the type of the reaction (i–iv) takes place. The above reasoning demonstrates why another approach is necessary, namely carrying out the entropy measurements.

An illustration of the concepts discussed earlier is shown in Fig. 2.2 where an electrode $Li_xV_2O_5 \mid Li^+$ reveals a number of features typical for EASP. The curve shows two areas where the OCV is almost independent upon composition with sharp transition from one domain to another. Two possible interpretations of this curve are

(a) the electrode behaves according to the mechanism iv, considered previously, or
(b) there exists a weak dependence of equilibrium potential on composition that has not been detected thus far.

2.3.4 Entropy of the Electrode Reaction

Solving a non-steady state problem of electrochemical kinetics means obtaining the relationships between current, potential, concentration, and time. Very often, these relationships can be found as the solutions of the differential equation known

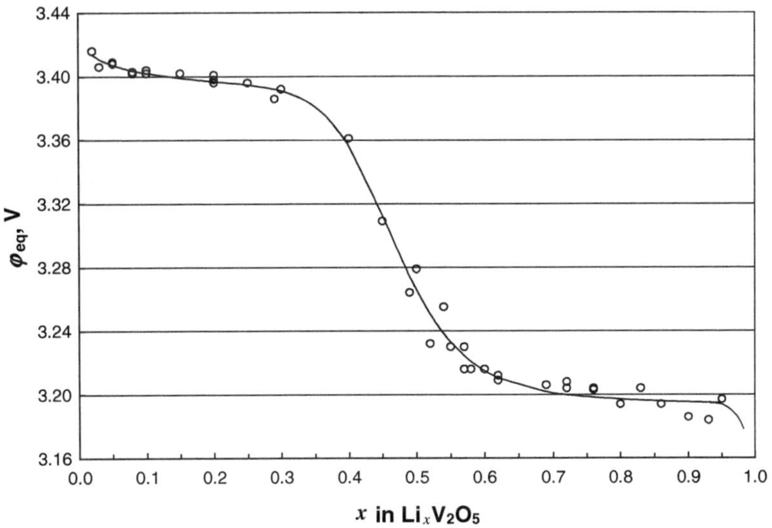

Fig. 2.2 OCV of the cell Li | $LiClO_4$ (propylene carbonate) | $Li_xV_2O_5$ at various x at 22 °C [11]

as second Fick's law with corresponding initial and boundary conditions. The latter are either: the equation $i = i(\varphi)$ or the relationship: $\varphi = \varphi(c)$, where i is the current density and φ is the electrode potential under load. The prior knowledge of the $\varphi = \varphi(c)$ dependence is presumably known from theory or preceding experiment.

The calculation approach presented here deviates from this direct scheme for the sake of the full consideration given to the entropy of electrode reaction and showing the importance of simultaneous scrutiny of entropy and Gibbs energy. Entropy of the reduction/oxidation (or discharging/charging) process is an effective measure of these structural changes, including the possible phase transitions. Statistical thermodynamics provides the direct connection between the entropy (S) and thermodynamic probability (w), which is the numeric measure of the disarrangement of the system, by Boltzmann's equation $w = \exp(S/k)$, where k is Boltzmann's constant. Crystalline lattice, being characterized by certain symmetry degree, must be sensitive to its ordering-disordering caused by electrochemical insertion or extraction of guest species. A redox-reaction occurring in an electrochemical cell gives a unique opportunity to directly measure the cell's Gibbs energy and entropy.

According to the Gibbs-Helmholtz equation

$$\Delta S = -zF\frac{E}{T} \tag{2.19}$$

where E is the emf of the electrochemical circuit, particularly, of Eq. 2.1. Thus, the emf is equal to the EP (φ) of the electrode under study in lithium scale (versus the Li | Li^+ reference).

Obviously, the entropy must depend on DoC of the electrode since the EP does:

$$\frac{\partial \Delta S}{\partial x} = -zF\frac{\partial(\phi/T)}{\partial x} = -zF\frac{1}{T}\frac{\partial \phi}{\partial x} \neq 0$$

Equation (2.15) with constant standard potential and (2.17) with the standard potential apparently depending on DoC can be rewritten in generalized form as

$$\phi = \phi^{\circ} + \Psi(T,x)$$

Equations (2.15) and (2.17) will then take the form

$$\phi(T,x) = \phi_1^{'}(T) + \Psi_1(T,x) \tag{2.20}$$

and

$$\phi(T,x) = \phi_2^{'}(T) + \Psi_1(T,x) \tag{2.21}$$

Then, it follows from Eq. 2.20 that

$$\frac{\partial \phi}{\partial T} = \frac{\partial}{\partial T}\phi_1^{\circ}(T) + \frac{\partial}{\partial T}\Psi_1(T,x)$$

or according to Eq. 2.19 (at $E = \varphi$),

$$\Delta S(T,x) = \Delta S_1^{\circ}(T) - zF\frac{\partial}{\partial T}\Psi_1(T,x)$$

Within a relatively narrow temperature range, far from melting point and far from the absolute zero, the derivative $\partial E/\partial T$ is constant, i.e., $\partial^2 E/\partial T^2 = \partial \Delta S/\partial T = 0$. If linear approximation is applied, $\partial \Delta S^{\circ}/\partial T = 0$ and:

$$\Delta S(x) = \Delta S_1^{\circ} - zF\frac{\partial}{\partial T}\Psi_1(T,x) \tag{2.22}$$

Therefore,

$$\frac{\partial \Delta S}{\partial x} = -zF\frac{\partial^2}{\partial T \partial x}\Psi_1(T,y) \neq 0$$

because $\partial \Delta S^{\circ}/\partial y = 0$ for the reaction (2.14).

In turn, dealing with the reaction (2.16), according to (2.21),

$$\frac{\partial \phi}{\partial T} = \frac{\partial}{\partial T}\phi_2^{\circ}(T) + \frac{\partial}{\partial T}\Psi_2(T,x)$$

In this case the result is

$$\Delta S(x) = \Delta S_2^{'} - zF\frac{\partial}{\partial T}\Psi_2(T,x) \tag{2.23}$$

Oppositely to Eq. 2.15, here $\partial \Delta S^{\circ}/\partial y \neq 0$, that is why

$$\frac{\partial \Delta S}{\partial x} = \frac{\partial}{\partial x}\Delta S_2^{\circ}(x) - zF\frac{\partial^2}{\partial T \partial x}\Psi_2(T,x) = \frac{\partial}{\partial x}[\Delta S_2^{\circ}(x) - zF\Psi_2(T,x)] \tag{2.24}$$

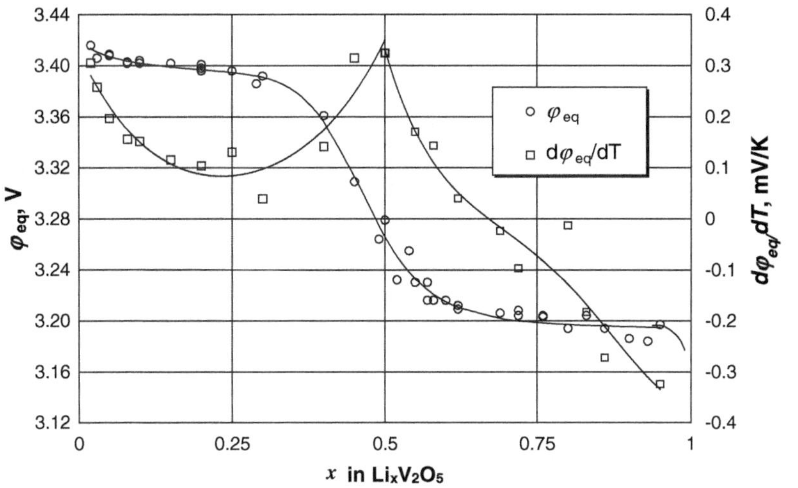

Fig. 2.3 OCV of the cell Li I $LiClO_4$ (propylene carbonate) I $Li_xV_2O_5$ at various x at 22 °C and its temperature coefficient dE/dT [11]

It follows from Eq. 2.24 that $\partial\Delta S/\partial x$ might be zero, in other words $\Delta S_2° - zF\Psi_2(T, x) = const$. This may mean that the difference in standard entropies, that is, the difference in structures of different compounds $Li_{p(x)}Z$ is entirely determined by lithium content, and no actual structural change occurred. Since determining of the *nature and composition* of the final product of the reaction can help to determine the *mechanism* of the reaction, the entropy calculations would be best supported by the structural and composition analyses.

In Fig. 2.3, the EP of $Li_xV_2O_5$ I Li^+ electrode and the electrode's temperature coefficient are overlapped in a single chart. Distinct dependence of entropy upon composition allows concluding that instead of the "true" plateaus on the equilibrium curve, a weak dependence upon composition takes place. A sharp peak at $x = 0.5$ most likely signals a phase transition at this composition.

2.4 Mass Transport in Solid Electrodes

Reduction (oxidation) of EASP begins with the enrichment (or depletion) of the surface layer with the product of the reaction (i.e., guest concentration change) followed by the guest distribution throughout the bulk of the material. The mechanism of distribution may vary, from the simple diffusion process to complicated multi-step solid-phase reactions including the formation of new phases. The following discussion is limited to the simple case of diffusive dissipation of inserted particles, and formation of the *phase of variable composition* (the solid solution).

For simplicity, the solid solution phase is considered nonporous and electronically conductive [34].[5,6] The latter assumption allows for absence of determination whether the guest entities are ions or atoms.

Diffusion occurring in solids during electrochemical processes never approaches steady state conditions. Therefore, only methods suitable for surveying of non-steady state electrochemical processes can be successfully employed for such investigations. The diffusion process follows second Fick's law

$$\frac{\partial C}{\partial t} = \nabla \cdot (D \nabla C) \tag{2.25}$$

or, for one-dimensional plane diffusion

$$\frac{\partial C}{\partial t} = \frac{\partial}{\partial x}\left(D\frac{\partial C}{\partial x}\right) \tag{2.25a}$$

where $D = D(C)$ is the diffusion coefficient (DC) *depending on concentration.* The flux density j is defined by first Fick's law.

$$j = -D\nabla C \tag{2.26}$$

Equation 2.26 is correct in any case where the spatial concentration difference is sufficient to drive the mass transport.

The dependence of DC upon concentration immensely affects the solution of the related equations, such as the Eq. 2.25. Equation 2.25 has the *analytical solution* if $D = const$, which is true for many border conditions realized in actual electrochemical experiments. However, if the DC changes with concentration, the equation can be solved only numerically. Therefore, it is as useful and convenient to conduct the experiments within such conditions, where the DC changes are negligible. Equation 2.25a is then reduced to

$$\frac{\partial C}{\partial t} = D\frac{\partial^2 C}{\partial x^2} \tag{2.25b}$$

The practical implementation of such measurements is challenging. First, the surface concentration perturbation occurring during the measurement must be as small as possible. However, the physical meaning of "as small as possible" is different for each material and experimental method. Lithium-ion EASPs are composed of lithium, transient metals, oxygen, and, possibly, other elements. Molar volume of these materials is typically fraction of liter per mole. For example, molar volume of lithium cobalt oxide, which is the main cathode material for

[5] The assumption of electrode materials' conductivity is an intentional oversimplification. In reality, vast majority of lithium-ion electrode materials possess rather low conductivity and neglecting their resistance is incorrect in terms of the strictly mathematic approach to balancing the electrode equations. However, this simplified calculating approach is used for clarity in presentation of a novel concept. For the complete mathematics of transport phenomena pertaining to the simpler electrochemical systems, the reader is directed to J. Newman's book [34].

[6] For the same reasons, only plane diffusion is considered here.

commercial LIC, is 19.45 cm^3 mol^{-1}. Lithium concentration in electrodes usually ranges up to few tens of moles per liter, rarely falling below several moles per liter. In Li_xCoO_2, the typical lithium concentration range is between ~20 and 51 mol L^{-1}; whereas in carbonaceous negative electrodes, the lowest lithium concentration approaches almost zero (at least at the surface), but also goes up to ~30 mol L^{-1} in the fully lithiated state. A liquid solution with such content of solute is considered a concentrated solution experiencing "strong deviations from ideality" of physical and electrochemical properties. In particular, the dependence of the DC on concentration takes the form of Nernst-Hartley equation [35]

$$D = D_0\left(1 + \frac{\partial \ln \gamma}{\partial \ln C}\right) \tag{2.27}$$

As demonstrated in previous sections, Eq. 2.27 is out of the direct use due to absence of the data on the activity coefficients γ in EASP. Instead of Eq. 2.27, another equation that contains DC is more helpful, namely the Einstein-Smoluchowski relation (Eq. 2.28) for mean square distance $\langle d^2 \rangle$ of a freely randomly walking particle during time t:

$$\left\langle d^2 \right\rangle = 2Dt \tag{2.28}$$

Equation (2.28) shows that any plane of constant concentration propagates proportionally to the square root of time. Based on the above considerations, Eq. 2.28 may serve the dual function: (a) it is useful for preliminary validation of the value of the experimentally obtained DC and (b) can be used for the preliminary estimation whether limited or semi-infinite diffusion model should be applied. These estimates are using the thickness of the sample and the experiment duration, as well as the predetermined possible range of DC values. For example, when experiment duration equals 1 h, a nonporous electrode with the thickness of few millimeters behaves as semi-infinite media when diffusion coefficients are greater than 10^{-16} cm^2 s^{-1}.

One of the most important methods of testing of the EASPs is their *galvanostatic reduction and oxidation*. The analogous method in electroanalytical chemistry is called chronopotentiometry (CPM). In CPM, the *transient time* corresponds to the *utilization* of the battery electrode. The transient time is the time when the surface concentration of the depolarizer drops to zero; the utilization is the fraction of inserted (extracted) lithium atoms. Obviously, at constant current conditions the difference between these characteristics is solely a matter of units. However, since the guest's concentration varies from (sometimes) zero to a certain maximal value, the DC will suffer all possible variations as well, and all the CPM equations cannot be used, except the single CPM relationship given below.

The theory of the CPM shows that the reversible curves obtained at different currents (i) obey the law: $i^2\tau = const$ where τ is the transient time. It has been shown with the help of the similarity theory that this rule covers any diffusion-controlled constant-flux process in semi-infinite media, and it works not only for transient time, but also for any time necessary to reach a certain potential regardless of the DC dependence upon concentration (or lack of thereof) [36].

Potentiodynamic reduction and oxidation (chronovoltammetry, CVA) is frequently (and sometimes, erroneously) employed for characterization of the processes on EASP with the accompanying formation of solid solution.[7] Very often, the equations for diffusion kinetics in semi-infinite media (Randles-Sevcik, Matsuda-Ayabe, and other [28] are employed for interpreting the results obtained via linear sweep of potential with the rate of 5–15 mV s^{-1} within the potential range of the electrode's OCP. However, these equations have been derived for constant DC, and do not apply in the above conditions. Even if the DC were constant, the approach is not relevant to the measurements associated with the EASP reduction, or lithium insertion. Indeed, *no Li-ion diffusion-related maximum appears* on the CVA-gram. The CVA current corresponding to insertion is an *infinitely increasing function*. The experimentally observed maxima are determined by other phenomena, different to that occurring in classic CVA (e.g., reaching the maximum of allowed guest's concentration, or new phase formation).

Regardless of the measurement technique, a common approach exists to address the lack of constant DC throughout the entire range of concentration (potential). It is possible to establish the intervals where the DC is constant. From the knowledge of the DC's dependence upon concentration within these intervals, the DC behavior for overall concentration range can be determined. Based on this approximation, an entire group of the applied electrochemical analyses methods have been enabled.

These electrochemical non-steady state methods are divided into three groups, where potential, current, or charge drive the process. The driving (controlling) factors in turn define the border and initial conditions for solving Eq. 2.25b [28]. The representatives of each group are described in the following section. All experiments must be performed on an electrode made of partially discharged EASP in lithium-ions containing solutions, initially at equilibrium potential. The initial concentration of lithium in the solid is denoted as c_0, and the current density i is used as the rate of the reaction, unless otherwise noted. The ionic transport in the solution is typically significantly faster than the diffusion within the solid, thus the electrolyte concentration can be assumed constant during the short-term experiment.

2.4.1 Galvanostatic Pulse Method

In the galvanostatic pulse method (GSPM) [37], lithium is inserted into (and extracted from) the surface layer of the electrode by a short rectangular current pulse i. The charge transfer step is presumed fast and the measured electrode potential is entirely determined by the surface concentration of lithium. During the

[7] The experiments with ultra-low scan rates (few microvolts per second) are beyond the scope of this paragraph but will be touched upon later, in Sect. 2.7.

experiment, the potential initially leaps to a certain maximal (minimal) value and then gradually returns to the initial value. The amount (in mol per area unit) of inserted lithium is $n = it/F$. Strictly speaking, the pulse should be instantaneous ($t \longrightarrow 0$). However, in practice, the sufficiently short pulse duration is less than the characteristic time of the diffusion.

The solution of Eq. 2.25b with initial $c(0, x) = c_0$ and border $(\partial c/\partial x)_{x=0} = 0$ (interface), $c(t,\infty) = c_0$ (bulk) conditions for $x = 0$ is [38]:

$$c(t, 0) = c_0 + \frac{i\tau}{F\sqrt{\pi D t}} = c_0 + At^{-1/2} \tag{2.29}$$

Here, τ is the pulse duration; the current i is positive for insertion and negative for extraction.

The DC can be calculated from the slope of the line illustrating the relationship of c *versus* $t^{-1/2}$. However, the observed value is the potential, which is a function of concentration $\varphi = \varphi(c)$. If the dependence of potential on concentration is of Nernst's type Eq. 2.29 takes the form

$$\exp\left[\frac{F}{RT}\Delta\phi(t)\right] = 1 + \frac{A}{c_0}t^{-1/2} \tag{2.30}$$

where $\Delta\varphi(t) = \varphi(t) - \varphi_{eq}$. For all EASP materials, the concentration must be expressed as a function of equilibrium potential.

The main disadvantage of the current pulse method lies in the ambiguous validity of the results. The ambiguity is caused by several factors. First, the requirement that $t \longrightarrow 0$ leads to the inversely proportional current increase, and may cause substantial concentration and/or the DC changes within the surface layer. These changes negate the application of Eq. 2.25b. Secondly, extending the pulse duration alters the experiment conditions while distorting the border condition of $x = 0$, thus rendering Eq. 2.29 inadequate. Appropriate balancing of these equations (Eqs. 2.25b and 2.29) needs the preliminary knowledge in the DC value. An estimate of the allowed pulse time can be obtained using Eq. 2.28. Another requirement is the shallow penetration of guests into the host lattice. If the penetration depth d corresponds to 2–3 atomic diameters, then for the ion O^{-2} with the ionic radius of (1.5–1.7) × 10^{-8} cm, [39] $d \approx 1$ nm. In this case, the allowed pulse duration is 0.05 and 500 ms for diffusion coefficient values between 10^{-10} and 10^{-14} $cm^2\ s^{-1}$. If lithium content change in such thin surface layer is 1 % (for example, in $LiCoO_2$ this corresponds to $\Delta c = 0.5$ mol L^{-1}). The resulting allowed pulse current density will fall between 100 and 10 μA cm^{-2}.

In materials with high diffusion coefficients ($D > 10^{-11}$ cm^2/s), the pulse current methods do not yield accurate results. This inaccuracy is caused by both technical and fundamental difficulties. The former problem is associated with the limited capabilities of the pulse generators. The latter problem is related to the speed of the double layer charging, where the fraction of 50 μs for the pulse duration might be not sufficient to achieve the double layer charging.

2.4.2 Potential Transient Method

In this method, [37] an electrode at equilibrium is polarized by a small shift of potential, thus keeping a constant concentration of inserted atoms on the surface of the electrode. The DC can be calculated from the registered current-versus-time curve called chronoamperogram. The method is also called single-step chronoamperometry (SSCAM). In this method, the condition $D \approx const$ is under better control than in the GSPM.

The solution of Eq. 2.24b for semi-infinite electrode at initial $c(0, x) = c_0$ and border $c(t, 0) = c_s$ and $c(t, \infty) = c_0$ conditions has the form

$$\frac{c - c_s}{c_0 - c_s} = \text{erf}\frac{x}{2\sqrt{Dt}} \tag{2.31}$$

while the current density

$$i = FD\left(\frac{\partial c}{\partial x}\right)_{x=0} = F(c_s - c_0)\sqrt{\frac{D}{\pi t}} \tag{2.32}$$

The height of the potential step is defined by the function $\varphi = \varphi(c)$. The preference for the low value of the potential step is related to the following causes: (a) the effect of the small potential step on the value of the DC is small, and (b) the dependence of the potential upon "guest" concentration can be simplified to linear relationship: $c_s - c_0 = \Delta\varphi$. Thus, the current decrease remains linear versus reciprocal square root of time ($t^{-1/2}$).

For the electrode of limited thickness δ, the border condition $c(t,\infty) = c_0$ must be substituted with $(\partial c/\partial x)_{x=l} = 0$ (impenetrable border). The expression for the current density then obtains a more complicated form, which can be found in a handbook dedicated to the mathematics of diffusion [39]. However, the sample can be considered as infinite if the dimensionless parameter $Dt/l^2 < 0.3$.

In such case, the calculation of DC is simple. The criterion for infinite sample dimensions yields that the largest possible $D = 10^{-8}$ cm^2/s. Thus, the electrode of 60 μm thickness behaves as semi-infinite during the 10^3 s current decay. Therefore, if the experiment's duration is limited to few minutes (10^2 s) then the allowed sample thickness would be ~20 μm. Such consideration is necessary when film or single-crystal electrodes are employed.

The examples of the described methods' applications can be found in the works by Nikolskaya et al. [14] where DC of lithium in niobium diselenide of layered structure $2H$-$NbSe_2$[8] has been determined by GSPM and SSCAM. For calculations,

[8] This material has the metallic type of conductivity and extremely flexible crystalline lattice (the lattice can expand along the c axis, doubling the parameter), and can be easily pressed into pellets. The degree of lithium-ion reduction is about 0.7 (i.e., lithium in the crystal is present in 70 % as atoms and 30 % as ions). Thus, layered niobium diselenide is the perfect material for testing the proposed models. In order to eliminate the pores, pellets have been impregnated with molten paraffin-polyethylene alloy under vacuum in all cited works.

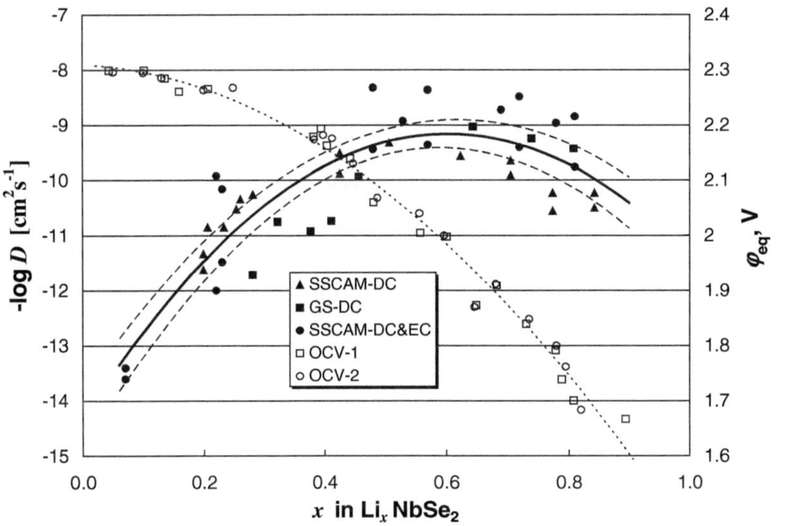

Fig. 2.4 OCV of the cell: Li I 1 M $LiClO_4$ (PC) I Li_xNbSe_2; DC of lithium at various x. The accuracy of the DC calculation was 0.2–0.3 in the order of magnitude (±1.6–2.0 in mantissa); *dashed lines* show 95 %—confidence interval. *References* OCV-17; OCV-2 [12]; SSCAM-DC and GS-DC [14]; SSCAM-DC and EC [15]

the equilibrium curve was obtained. The experimental results are well consistent between various measurement methods (Fig. 2.4). The DC varies from 10^{-11} to 10^{-9} $cm^2\ s^{-1}$ passing over a maximum at $x \approx 0.5$–0.6. This is the region where the interlayer distance reaches the maximum [40] not changing with further lithium content increase. However, the number of available diffusion leaps decreases, which leads to the DC decrease.

The DC calculations did not include any assumption of the concentration dependence of the DC. Comparing the DC and OCV curves, it is easy to ascertain that there is no direct proportion between $f_1 = \frac{\partial \varphi(x)}{\partial \ln x}$ and $f_2 = D(x)$, which should exist due to Eq. 2.27. Thus, from Nernst's equation:

$$\phi = \text{const} - \frac{RT}{F} \ln \gamma c$$

where γ is the activity coefficient, it follows that

$$\frac{\partial \ln \gamma}{\partial \ln c} = \frac{F}{RT} \frac{\partial \phi}{\partial \ln c} - 1$$

The above constitutes another reason for avoiding the term of activity in studying the EASP.

2.5 Rate of Electrochemical Stage

The rate i of the electrochemical reaction depends on reactants' concentrations c, temperature T, and electrode potential φ. For mono-molecular process,

$$i = k(T, \phi)c \tag{2.33}$$

where $k(T,\varphi)$ is the rate constant. Some authors prefer using thermodynamic activity instead of concentration in the equations of the same type as Eq. 2.33. Previous sections of this text (Part 2.1) demonstrated that there was no viable reason for involving the concept of activity in thermodynamic considerations of EASP electrodes. Similarly, there are no reasons to use the activity in kinetic reactions [41].[9]

The necessary but not exclusive condition for the reaction to occur is the collision of particles, or the act of particles hitting the surface of another phase. Equation 2.33 reflects the assertion that the rate is directly proportional to the frequency of collisions. For a decomposition reaction, the number of bond breakages per unit of time is proportional to total number of particles.

However, not every collision leads to the product formation since many collisions are elastic. The collective efficiency of collisions is described by the reaction's rate constant. The rate constant depends on temperature and potential and on various environment factors, such as the *nature* of the solvent and electrolyte, and the *quantitative composition* of electrolyte solution.

2.6 Concentration Dependence of Exchange Current on EASP

In order to perform the relevant kinetic calculations, first, a specific mechanism of the reaction must be presumed. Thus, the process rate (expressed as current density[10]) of a single-step one-electron redox-reaction:

$$\mathrm{Ox} + ne \leftrightarrows \mathrm{Red} \tag{2.34}$$

obeys the theory of mixed kinetics when both charge transfer and diffusion control the overall rate of the reaction (Butler-Volmer equation, BVE):

$$i = i_0 \left[\frac{c^{\mathrm{S}}_{\mathrm{Ox}}}{c_{\mathrm{Ox}}} \exp(-\alpha n \vartheta \eta) - \frac{c^{\mathrm{S}}_{\mathrm{Red}}}{c_{\mathrm{Red}}} \exp(\beta n \vartheta \eta) \right] \tag{2.35}$$

[9] In the absolute reaction rate theory, the rate of the reaction is proportional to the concentration of activated complex, which is in equilibrium with reactants. Thus, the activated complex concentration can be calculated using the corresponding equilibrium constant and, consequently, the activities. This is the reason why the activity coefficients appear in kinetic equations.

[10] Unless otherwise noted, the term "current" is equivalent to the "current density" in the following text.

where i_0 is the exchange current, α and β are cathode and anode transfer coefficients ($\alpha + \beta = 1$), $\eta = \varphi - \varphi_{eq}$—polarization, $\vartheta = F/RT$. The index "s" marks surface concentrations of both forms of the depolarizer. In fact, the observed current is the difference $i = \vec{i} - \overleftarrow{i}$ between the currents of direct $\vec{i}$ and reverse $\overleftarrow{i}$ processes of (2.24). The formulas for partial currents $\vec{i} = nFk^{\circ}c_{Ox}\exp[-\alpha n\vartheta(\phi - \phi^{\circ})]$ and $\overleftarrow{i} = nFk^{\circ}c_{Ox}\exp[\beta n\vartheta(\phi - \phi^{\circ})]$ include standard rate constant k°, and the potential is counted from the standard potential value (φ°) for the reaction (Eq. 2.34). The dependence of the exchange current on concentration is usually expressed as

$$i_0 = i_0^o\, c_{Ox}^{\beta}\, c_{\text{Red}}^{\alpha}, \tag{2.36}$$

where i_0^{o} is the *standard exchange current* having the following dimensionality: (current) × (length)/(matter amount), e.g., A cm mol^{-1}. However, the formula (Eq. 2.36) is applicable only in case of the Nernst-type potential behavior being derived from the equation

$$i_0 = nFk^{\circ}c_{\text{Ox}}\exp[-\alpha n\vartheta(\phi_{eq} - \phi^{\circ})] = nFk^{\circ}c_{\text{Red}}\exp[(\beta n\vartheta(\phi_{eq} - \phi^{\circ})] \tag{2.37}$$

when $\phi = \phi^{\circ} + \frac{RT}{nF}\ln\frac{c_{\text{Ox}}}{c_{\text{Red}}}$. Then $i_0^{\text{o}} = Fk^{\circ}c_{\text{Ox}}\exp\left(-\alpha\frac{nF\phi^{\circ}}{RT}\right)$

If $\varphi = \varphi(c)$ is not expressed by the Nernst equation, then $\vartheta \neq F/RT$. Indeed, at equilibrium ($\varphi = \varphi_{\text{eq}}$) the currents of direct and reverse reactions are equal, $i = 0$, and

$$c_{\text{Ox}}\exp[-\alpha nf(\phi_{eq} - \phi^{\circ})] = c_{\text{Red}}\exp[\beta n\vartheta(\phi_{eq} - \phi^{\circ})] \tag{2.38}$$

In logarithmic form (Eq. 2.38)

$$\phi_{eq} = \phi^{\circ} + \frac{1}{(\alpha + \beta)n\vartheta}\ln\frac{c_{\text{Ox}}}{c_{\text{Red}}} \tag{2.38a}$$

if $\vartheta = F/RT$, then the entire relationship (Eq. 2.37) can be reduced to the form equivalent to the Nernst's Eq. (2.38a).

It follows from Eq. (2.36) that increasing of concentration of both oxidized and reduced forms leads to the increase in exchange current. Oppositely, taking logarithm and the derivative in respect to c_{Red} of the second equation (Eq. 2.37) the following can be derived:

$$\frac{d\ln i_0}{d\ln c_{\text{Red}}} = 1 + \beta\phi_{eq}\frac{d\vartheta}{d\ln c_{\text{Red}}} \tag{2.39}$$

since $= 1/|\partial\varphi_{eq}/\partial\ln c)|$.

It follows from (Eq. 2.30) that the exchange current increases if $\frac{d\phi_{eq}}{d\ln c_{\text{Red}}} > -\frac{RT}{\beta nF}$, and decreases in the opposite case.

In the phenomenological theory the transfer coefficient is an empirical quantity, which is usually equal to 0.5, and is not dependent on the concentration and the potential (the Brønsted-Polanyi-Semenov principle). Therefore, experimental testing of the proposed approach should provide an answer, in particular, to the question of whether the principle is obeyed in this case.

In physical sense, the substitution of F/RT with $1/|\partial\varphi_{eq}/\partial\ln c)|$ is relative to the logical device realized in the VTF model of the conductivity (see Chap. 3) where, in fact, the absolute zero temperature is substituted with a sort of "relative zero temperature": the T_0 temperature (Eq. 3.3) is the temperature at which the ions mobility ceases to exist. Similarly, the value of $1/|\partial\varphi_{eq}/\partial \ln c)|$ can be interpreted as the "$F/r(c)T$" where the absolute gas constant R is substituted with a "relative gas constant r" depending on concentration c and is specific to a given material.

2.7 The Methods of Investigation of Electrochemical Reaction Step

The step of charge transfer in the multi-step process of electrochemical reduction and oxidation of EASP is relatively fast. Reliable data on the kinetic parameters of this step (exchange current or rate constant and transfer coefficient) can be established by using high-frequency methods examining short-term response of the electrode, with experiment times being much shorter than these allowed for studying of the diffusion step. The requirements for minimal perturbation are the same as those recommended for diffusion investigations. However, at early stages of electrode disturbance, the double-layer capacity (C) can affect the results, and together with high resistance of the EASP electrodes (R), lead to high value of the *time constant* (RC) of the electrochemical cell. The high value of RC is capable of distorting the high-frequency experimental results.

Many electrochemical analytical *combined methods* allow for simultaneous calculation of DC, exchange current (EC), and transfer coefficients. For example, in a method called the coulostatic or charge-step method [28] a current pulse of 0.1–1 μs is applied to the electrode, and the variation of the electrode potential with time after the pulse (that is, at *open circuit*) is recorded. The method is very similar to GSPM; however, the instrumentation is different, namely, the charge is injected by discharging a small capacitor across the electrode. The results are not affected by the double-layer capacity and electrode resistance.

The SSCAM is the most advantageous method due to the reasons given in Part 2.2. However, to get consistency of the DC values calculated from the results of combined methods and from the experiments where the diffusion exclusively controls the overall rate, the experimental values of $\vartheta = d\varphi_{eq}/d\ln c$ must substitute the $\vartheta = nF/RT$ in the BVE.

The equation for mixed kinetics in SSCA, has the form [42]

$$i(t) = I(0)\exp(\lambda^2 t)\mathrm{erfc}(\lambda\sqrt{t}) \tag{2.40}$$

where $I(0)$ and λ are the formal parameters

$$I(0) = i_0\left[\exp\left(-\alpha\vartheta\eta\right) - \exp\left(\beta\vartheta\eta\right)\right] \tag{2.41}$$

and

$$\lambda = i_0\frac{\exp(\beta n\vartheta\eta)}{nFc_0\sqrt{D}}. \tag{2.42}$$

For simplicity, the latter equation neglects the diffusion limitations in the liquid phase. The two limiting cases, when t → 0 and t → ∞, yield from Eq. 2.30:

$$i(0) = I(0)$$

which corresponds to the pure sluggish charge transfer reaction, and

$$i(\infty) = \frac{I}{\lambda}\frac{1}{\sqrt{\pi t}}$$

This is, in fact, Eq. 2.31.

Toroshchina et al. have conducted the SSCAM experiments with Li_xNbSe_2 and $Li_xV_2O_5$ [15]. Small potential signal (±4–12 mV) was applied to the electrode L_xZ | 1 M $LiClO_4$ (PC) at equilibrium at various degree of discharge, and the response chronoamperogram was registered during 100–200 ms. The formal kinetic parameters $I(0)$ and λ have been determined by fitting the experimental curves to Eq. 2.40 by the nonlinear least-squares method, and the real parameters have been calculated from Eqs. 2.41 and 2.42.

The obtained DC values, presented in Fig. 2.4, labeled as "SSCAM-DC&EC" demonstrate good agreement between the results obtained by different methods.

To find the EC and transfer coefficient, Eq. 2.41 has been converted into two convenient forms proposed by Essin in 1940 [42].

$$\log J_+ \equiv \frac{i}{1 - \exp(\vartheta\eta/RT)} = \log i_0 - \alpha\vartheta\eta$$

and

$$\log J_- \equiv \frac{i}{\exp(-\vartheta\eta/RT) - 1} = \log i_0 + \beta\vartheta\eta$$

Both log J_+ and log J_- linearly depend upon polarization η, and the both transfer coefficients can be calculated from the slope of the straight lines (Fig. 2.5). The experimental value of ϑ has been used in all calculations. The relation $\partial\eta/\partial\ln J_\pm = 2(\partial E_{eq}/\partial\ln c_{Red})$ means that $\alpha = \beta = 0.5$ is satisfactorily fulfilled and does not depend on x.

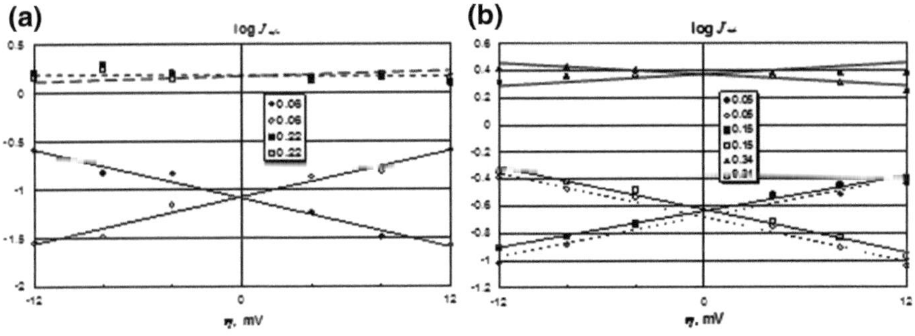

Fig. 2.5 Polarization curves in Essin's coordinate set for Li_xNbSe_2 (**a**) and $Li_xV_2O_5$ (**b**)

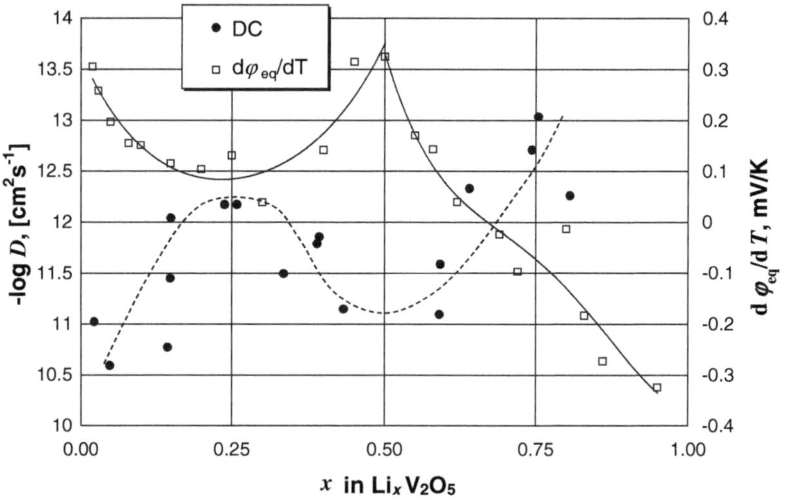

Fig. 2.6 Diffusion coefficient of lithium in LixV$_2$O$_5$ at various temperatures, and entropy of the reaction of insertion

The composition dependence of the diffusion coefficient of lithium in the lithium vanadate $Li_xV_2O_5$ has a complicated character (Fig. 2.6). The extremes on the DC—*x* the entropy—*x* curves appear at the same values of *x,* which is not surprising since both parameters are connected to the material structure. Calculation of the DC from SSCAM experiment using $\vartheta = nF/RT$ did not reveal such correlation indirectly confirming the rationality of the developing approach.

In contrast to the case of the diffusion coefficients, a direct relationship with the structure is not that obvious for the exchange current. However, the calculations also reveal a relationship between the course of the entropy versus composition plots and the exchange current versus composition plots (Fig. 2.7): in the region where one phase exists (region $x < 0.5$ or $x > 0.5$ for $Li_xV_2O_5$, and the entire range of *x* for Li_xNbSe_2) the curves are monotonic, and phase transformations cause changes in their character. The results shown in Figs. 2.6 and 2.7 stand in contrast

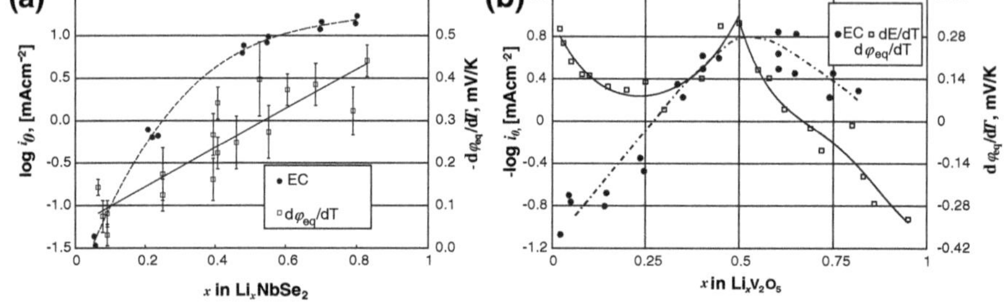

Fig. 2.7 Exchange current at various x on Li_xNbSe_2 (**a**) and $Li_xV_2O_5$ (**b**)

to some published work [43–46], where calculations were performed without consideration of the *experimental dependence* of the equilibrium potential on x. Such approach did not reveal dependence of i_0 on the composition of the solid phase. In particular, the authors of the work [43] have not found any effect of lithium content in Li_xNbSe_2 on exchange current, despite such effect being in existence.

2.8 Selected Examples of OCP and Related Dependencies on Lithium Content

In the previous parts of this chapter, few typical OCV curves have been described from the phenomenological point of view. Besides the curves presented here, numerous measurements of the OCP have been published for various materials; thus covering the full spectrum of possible types of equilibrium potential versus composition curves.

J.D. Raistrick et al. [2, 6] were perhaps the first who studied the thermodynamics of electrochemical insertion of lithium into EASP. In these works, Gibbs energy and entropy of lithiation of tungsten and vanadium bronzes have been measured along with the determination of lithium diffusion coefficient.[11] The OCP and entropy curves of $Li_xNa_{0.56}WO_3$ were smooth, confirming formation of the solid solution within the interval $0.0004 < x < 0.35$. However, $d\varphi_{eq}/dx \neq d\Delta S/dx$, similarly to the case of Li_xNbSe_2 where the OCP curve can be approximated by a parabolic curve (Fig. 2.4), while the $\Delta S - x$ dependence is linear (Fig. 2.7a).

Figure 2.8 represents the OCV and its temperature coefficient corresponding to the process of lithium insertion into electrochemically deposited manganese (IV) oxide [10]. In the area $x < 0.7$, the OCP demonstrates strong dependence upon lithium content, while the temperature coefficient remains constant. The interpretation of such peculiar phenomenon follows the case iii mechanism in Part 2.2.3.

Formation of a stage structure during lithium intercalation into graphite is a commonly recognized phenomenon and was touched upon in Chap. 1 of this book, in the section dedicated to carbonaceous anodes [47]. The OCP and entropy diagrams presented in Fig. 2.9 are perfectly consistent with the staging model. The areas where both entropy and Gibbs energy do, or do not depend upon lithium concentration, coincide with the corresponding composition ranges of different phases (stages).

Figure 2.10 presents the thermodynamic and crystallographic properties of lithium nickel cobalt oxide $Li_xNi_{0.8}Co_{0.2}O_2$.[12] The entropy curve is consistent with

[11] The diffusion coefficients found in the $Li_xNa_{0.56}WO_3$ bronzes are of the order of magnitude from 10^{-9} to 10^{-7} $cm^2\ s^{-1}$. Such a high values of the DC would eliminate the rate-capability limitations of the cathode material. So far, however, the high-rate LIB with tungsten bronze-based anode has not been reported.

[12] Mars rovers "Spirit" and "Opportunity" as well as other speciality battery consumers employ lithium-ion batteries with cathodes based on this material.

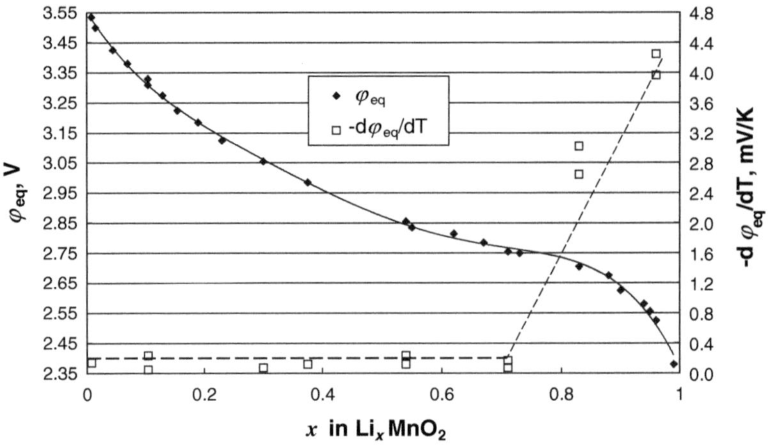

Fig. 2.8 OCV of the cell Li | 1 M $LiClO_4$ (PC) | Li_xMnO_2 at various x at 20 °C and its temperature coefficient dφ_{eq}/dT

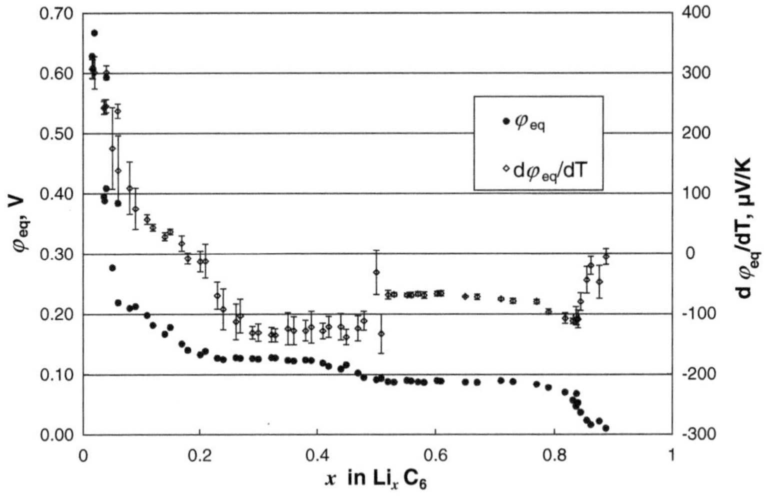

Fig. 2.9 OCV of the cell Li | 1 M $LiPF_6$ (carbonates blend) | Li_xC_6 at various x at 20 °C and its temperature coefficient dφ_{eq}/dT

the direct crystallographic measurements. Thus, three areas of the curve can be highlighted. There is a peak at $x \approx 0.85$, and a plateau at approximately $0.4 < x < 0.55$ where $\Delta S \approx 0$. The peak indicates that a phase transition develops in the concerned composition region, while in the plateau area no changes in the crystalline lattice are observed. At $0.2 < x < 0.3$, the entropy slightly increases and remains constant at $x < 0.3$. From this approach, it follows that the cathode

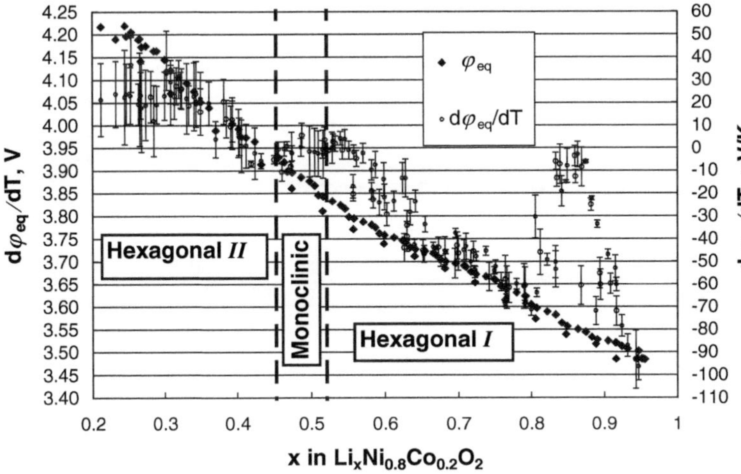

Fig. 2.10 OCV of the cell Li | 1 M $LiPF_6$ (carbonates blend) | $Li_xNi_{0.8}Co_{0.2}O_2$ at various x at 20 °C and its temperature coefficient $d\varphi_{eq}/dT$; [48] areas of different crystalline structures determined by XRD are indicated. Unpublished Yardney internal R&D project

material experiences the greatest impact during the last period of the charge where the electrode potential drops below ~3.6 V. From these measurements it is apparent that the smallest effect on the crystalline lattice is observed when the potential of cathode material changes from ~3.8 to 4.0 V, i.e., x in $Li_xNi_{0.8}Co_{0.2}O_2$ varies from ~0.35 to 0.55. In this region, the electrode delivers about 25–30 % of the full capacity. The practical confirmation of the OCP curves was done in large, manufacturing-type cells (25 Ah) developed for an orbiter application by Yardney Technical Products, Inc. The cells were charged to 3.9 V and discharged at C/2 rate during 1 h. After 25,000–30,000 cycles, their 100 % DOD capacity was 87–90 % of the initial.

Finally, with Fig. 2.11 taken as an example, the method called "dynamic capacity" is discussed [49] When employing the methods of investigation of the EASP, some researchers use the CVA at extremely low scan rates, down to few microvolts per second. It is suggested that at such rate, no concentration gradient forms in the solid material, and the insertion proceeds at quasi-equilibrium conditions. The value i/v is called the dynamic capacity (v is the potential scan rate), and is equivalent to the derivative $dx/d\varphi_{eq}$. It is believed that the analysis of the $dx/d\varphi_{eq}$ versus φ_{eq} allows determining the points of phase transitions.

Overlapping of two curves $dx/d\varphi_{eq}$ and $d\varphi_{eq}/dT$ demonstrates that two extremes on the former curve (at $\varphi_{eq} \approx 3.55$ and 3.60 V) match the peaks on the entropy curve. However, the well-shaped peak at $\varphi_{eq} \approx 3.70$ V on the $dx/d\varphi_{eq} - \varphi_{eq}$ curve most likely *does not* correspond to a phase transition. Further, there is no indication of possible phase transitions at $\varphi_{eq} > 3.9$ V, which is prominent on the $d\varphi_{eq}/dT{-}\varphi_{eq}$ curve.

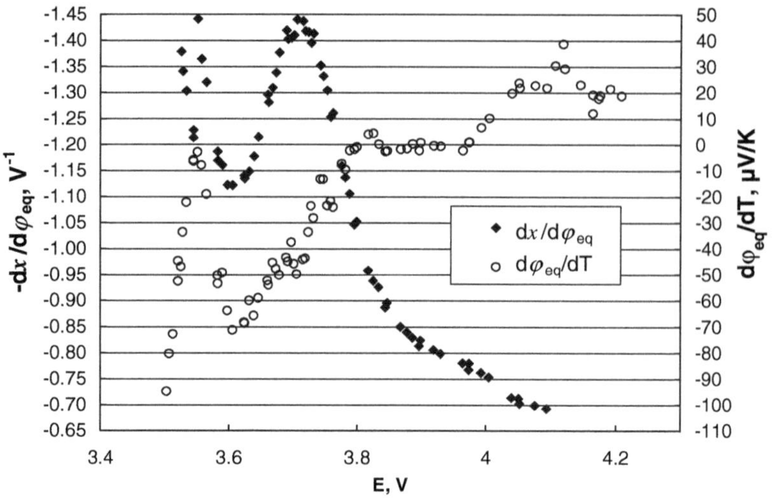

Fig. 2.11 $dx/d\varphi_{eq}$ and $d\varphi_{eq}/dT$ versus OCV for the cell Li | 1 M $LiPF_6$ (carbonates blend) | $Li_xNi_{0.8}Co_{0.2}O_2$ at various x at 20 °C

2.9 Conclusion

Modeling of the processes occurring in lithium ion batteries during cycling is impossible without knowledge of fundamental thermodynamic and kinetic properties of the electrode materials. The chapter offered an approach for interpreting the dependencies of thermodynamic functions, and determining kinetic parameters of the reaction of electrochemical insertion–extraction of foreign ions into solid materials.

The chapter intentionally avoided appealing to the detailed microscopic description of the materials and mechanisms of the reactions trying to show the self-sufficiency of pure electrochemical methods for achieving the goals defined as collection of the data necessary for modeling lithium-ion batteries.

Acknowledgments The author is solely responsible for all errors, inconsistencies, misinterpretations, misconceptions, omissions, etc. The author expresses his greatest and warmest gratitude to his teachers: Prof. Adolph A. Ravdel and Prof. Konstantin I. Tikhonov, and to his friends and colleagues: (in alphabetic order) Dr. B.G. Karbassov, Dr. E.Yu. Nikolskaya, Dr. M.Yu. Pozin, Dr. S.I. Shustova, Dr. I.A. Srago, Dr. E.I. Toroshchina, Ms. S.A. Trebukhova, and Dr. E.G. Vinogradova-Volzhinskaya whose great contribution made this review possible.

References

1. Modified after Encyclopaedia Britannica. Accessed on July 17, 2013, http://www.britannica.com/EBchecked/topic/264278/heterogeneous-reaction
2. Raistrick ID, Mark AJ, Huggins RA (1981) Thermodynamics and Kinetics of the Electrochemical Insertion of Lithium into Tungsten Bronzes. Solid State Ionics, 5:351–354. doi: 10.1016/0167-2738(81)90265-4

3. Jacobsen T, West K, Atlung S (1982) Electrostatic Interactions during the Intercalation of Li in Li_xTiS_2. Electrochim Acta 27:1007–1011. doi: 10.1016/0013-4686(82)80102-3
4. Dahn DC, Haering RR (1982) Phase Mixtures and Staging in Intercalated Li_xNbSe_2. Solid State Comm 44:29–32. doi: 10.1016/0038-1098(82)90706-2
5. Green M, Kang K (1983) Sodium Tungsten Bronze Thin-Films—Variation of Chemical-Potential with Sodium Concentration. Solid State Ionics, 8:281–289. doi: 10.1016/0167-2738(83)90002-4
6. Raistrick ID (1983) Lithium Insertion Reactions in Tungsten and Vanadium Bronzes. Solid State Ionics 9:425–430. doi: 10.1016/0167-2738(83)90270-9
7. Nikolskaya EY, Tikhonov KI, Rotinyan AL (1984) Nature of the Open-Circuit Potential of Li^+/Li_xNbSe_2. Sov Electrochem 20:237–240
8. Tye FL (1985) Manganese-Dioxide Electrode - X. A Theoretical Treatment based on the Concept of Two Solid Solutions in the Range gamma-MnO2 to delta-MnOOH. Electrochim Acta 30:17–23. doi: 10.1016/0013-4686(85)80053-0
9. Shustova SI, Ravdel BA, Tikhonov KI. (1985) Electrode-Reactions in the Reduction of Iron Sulfides in Propylene Carbonate. Sov Electrochem 21:524–527
10. Ravdel BA, Pozin MY, Tikhonov KI, Rotinyan AL (1987) Thermodynamic Properties of the Electrochemical Cell: Li I $LiClO_4$ (Propylene Carbonate) I Li_xMnO_2. Sov Electrochem 23:1459–1464
11. Toroshchina EI, Ravdel BA, Tikhonov KI (1987) Thermodynamic Properties of the Electrochemical-Cell Li/$LiCiO_4$ (Propylene Carbonate)/$Li_xV_2O_5$. Sov Electrochem 23:1435–1438
12. Toroshchina EI, Ravdel BA, Tikhonov KI (1991) On the Phase-Transitions and Phase-Composition in Cathode Reduction of 2H-Niobium Diselenide. Solid State Ionics 48:267–269. doi: 10.1016/0167-2738(91)90041-9
13. Gavrilyuk VI, Plakhotnik VN (1994) Thermodynamics of Compounds of Lithium Intercalation into Vanadium Pentoxide. Russ J Phys Chem 68:1373-1376
14. Nikolskaya EY, Tikhonov KI, Rotinyan AL, Ravdel BA (1988) Diffusion-Coefficient of Lithium Ions as a Function of Lithium Concentration in Li_xNbSe_2. Sov Electrochem 24:495–498
15. Toroshchina EI, Ravdel BA, Tikhonov KI (1990) Features of the Calculation of the Kinetic-Parameters of Electrochemical Reactions on Solid Cathode Materials. 2. Sov Electrochem 26:611–617
16. Ravdel BA (1994) Kinetics of Lithium Nickel Cobalt Oxide Electrode Material for Lithium-ion Batteries. Proceedings of the 186th Meeting of the Electrochem Soc. Miami Beach, Florida, Oct. 9–14, 1994. Abstract #638
17. Newman J, Thomas KE (2002) Proceedings of the International Meeting on Lithium Batteries, IMLB'11, Monterey, CA, June 23–28, 2002. Abstract #346
18. Yazami R, Reyner Y, Fultz B (2002) Proceedings of the International Meeting on Lithium Batteries, IMLB'11, Monterey, CA, June 23–28, 2002. Abstract #225
19. Thomas KE, Newman J (2003) Thermal Modeling of Porous Insertion Electrodes. J Electrochem Soc 150:A176–A192. doi: 10.1149/1.1531194
20. Fultz B, Reyner Y, Swan-Wood T, Graetz J, Rez P, Ozawa Y, Lam K, Yazami R (2004) Proceedings of the 204th Meeting of The Electrochem Soc. Orlando, Florida, Oct. 12–16, 2004. Abstract #327
21. Reyner YF, Fultz B, Yazami R (2004) Proceedings of the 204th Meeting of The Electrochem Soc. Orlando, Florida, Oct. 12–16, 2004. Abstract #1361
22. Ravdel BA, Trebukhova S, Puglia FJ (2009) Thermodynamics of Electrode Materials for Lithium-Ion Batteries. Proceedings of the 216th Meeting of the Electrochem Soc. Vienna, Austria, Oct. 4–9, 2009. Abstract #640
23. Zhang Q, Guo Q, White RE (2006) A New Kinetic Equation for Intercalation Electrodes. J Electrochem Soc 153:A301–A309. doi: 10.1149/1.21422874
24. Bazant MZ (2013) Theory of Chemical Kinetics and Charge Transfer based on Nonequilibrium Thermodynamics. Acc Chem Res 46:1144–1160. doi: 10.1021/ar300145c
25. International Union of Pure and Applied Chemistry; http://www.iupac.org/. Accessed on July 19, 2013

26. Atkins PW (1994) Physical Chemistry 5th Edition. Oxford University Press, Oxford, England
27. Lewis GN (1901) The Law of Physico-Chemical Change. Proc American Acad Arts and Sci 37:47–69
28. Bard AJ, Faulkner LR (2001) Electrochemical Methods. Fundamentals and Applications 2nd Edition. John Wiley & Sons, Inc., Hoboken, New Jersey
29. Hansen M, Anderko K (1958) Constitution of Binary Alloys. McGraw-Hill, New York
30. Nikolskaya EY, Tikhonov KI, Semenovkobzar AA, Yanaki AA, Rotinyan AL (1981) Electrochemical Reduction and Oxidation of Niobium Diselenide in propylene carbonate. J Appl Chem-USSR 54:742–745
31. Nikolskaya EY, Tikhonov KI (1981) The Effects of Solvents and Electrolytes on Electrochemical Niobium Diselenide Reduction. Sov Electrochem 17:1079–1083
32. Leden I (1964) On the Thermochemistry of Stepwise Complex Formation. Pure Appl Chem 8:171–178. doi: 10.1351/pac196408020171
33. Vinogradova-Volzhinskaya EG, Ravdel BA, Tikhonov KI, Kozhevniko PS, Oshe EK, Nadezhin YS (1988) Electrochemical Reduction of Copper(II) Oxide Films in Propylene Carbonate. Russ J Electrochem 24:630–634
34. Newman J, Thomas-Alyea KE (2004) Electrochemical Systems 3rd Edition. John Wiley & Sons, Inc., Hoboken, New Jersey
35. Bockris JO'M, Reddy AKN (1998) Modern Electrochemistry, Vol. 1: Ionics 2nd Edition. Kluwer Academic/Plenum Publishers. New York, NY
36. Ravdel BA, Sheinin DA, Tikhonov KI (1987) Criterion for a Diffusion Mechanism of Processes Attended by Ion Incorporation into a Solid Electrode. Sov Electrochem 23:652–652
37. Kuta J, Yeager E (1977) Overpotential Measurements. In: Techniques of Electrochemistry, Vol. Measurements pp. 141–292, Yeager E, Salkind AJ, eds. Wiley-Interscience, New York, NY
38. Crank J (1975) The Mathematics of Diffusion. Oxford University Press, Oxford, England
39. Lide DR, ed (2001) CRC Handbook of Chemistry and Physics 82nd Edition. CRC Press, Boca Raton
40. Dahn D.C., Haering R.R. (1982) Solid State Comm., 44:29–33
41. Glasstone S, Laidler KJ, Eyring H (1941) The Theory of Rate Processes, McGraw-Hill Book Co., Inc., New York, NY
42. Vetter K (1967) Electrochemical Kinetics, Academic Press, New York, NY
43. Nikolskaya EY, Miznenko NN, Tikhonov KI (1988) Rate of the Step of Lithium Discharge and Ionization at Li_xNbSe_2 Intercalate. Sov Electrochem 24:489–494
44. Munshi MZA, Smyrl WH (1991) Insertion Reactions of Sodium in V_6O_{13} Single Crystals from a Solid Polymeric Electrolyte. Solid State Ionics 45:183–189. doi: 10.1016/0167-2738(91)90151-Z
45. Chen L, Vanzomeren A, Schoonman J (1992) Electrochemical Insertion of Lithium into $YBa_2Cu_3O_7$-Y Solid State Ionics 50:55–58. doi:10.1016/0167-2738(92)90036-O
46. Ratnakumar BV, Nagasubramanian G, DiStefano S, Bankston CP (1992) Kinetics of Intercalation of Lithium into $NbSe_3$ and TiS_2 Cathodes. J Electrochem Soc 139:1513–1521
47. Ogumi Z, Inaba M (2002) Carbon Anodes. Advances in Lithium-Ion Batteries. van Schalkwijk WA, Scrosatti B, eds. Kluwer Academic/Plenum Publishers, New York, NY
48. Ravdel B, Trebukhova SA, DiCarlo JF (2006) Stability of Lithium-ion Cells Evaluated from Thermodynamic Measurements. Proceedings of the 42nd Power Sources Conference, June 12–15, 2006, Philadelphia, PA, Abstract #9.3
49. West K (1989) Intercalation Compounds: Metal Ions in Chalcogenide and Oxide Hosts. In: High Conductivity Solid Ionic Conductors: Recent Trends and Applications. T.Takahashi, ed

Chapter 3
Optimizing Electrodes for Lithium-ion Cells

Stuart G. Santee, Boris Ravdel, Malgorzata K. Gulbinska, Joseph S. Gnanaraj and Joseph F. DiCarlo

Abstract Selection of active electrode materials is only the first step in the cell design and manufacturing process. In order to achieve the best cell performance, the "non-active" components need to be carefully selected both qualitatively and quantitatively. Especially in high-end applications, the electrode formulation optimization process is nontrivial and seriously impacts the cell's performance. Achieving the high performance limits, such as extremely high power, very high temperatures, or very long life, require thoughtful and deliberate approach to electrode formulation optimization.

3.1 Introduction

When lithium-ion cells were first commercialized in the early 1990s, the application target for these cells was portable electronics. As such, a size was chosen, and almost immediately after this product was launched, optimization work has begun to increase energy density and cycle life. Since that time, lithium-ion technology use has proliferated into many different applications such as power tools, aerospace, transportation, and medical devices. It is obvious that each application would want to exploit different attributes of this technology such as power density, cycle life, and energy density, not to mention cost. Exploitation of these attributes begins with material selection but then the majority of the enhancement comes from optimization of all subcomponents, namely cathode, anode, electrolyte, and separator. This first optimization of the individual cell's subcomponents is usually followed by the second, systemic optimization

S.G. Santee · B. Ravdel · M.K. Gulbinska · J.S. Gnanaraj (✉)
Yardney Technical Products, Inc, 2000 South County Trail,
East Greenwich, RI 02818-1530, USA
e-mail: joeg@yardney.com

J.F. DiCarlo
BASF Corporation, 25 Middlesex/Essex Turnpike, Iselin, NJ 08830-0770, USA

M.K. Gulbinska (ed.), *Lithium-ion Battery Materials and Engineering*,
Green Energy and Technology, DOI 10.1007/978-1-4471-6548-4_3

of the component formulations and cell design where the mutual relationships between the components (e.g., the relationship between the conductive carbons and binder content) are being taken into consideration. In the end, a highly customized set of lithium-ion cell designs is developed, whose suite of proprieties ranges from resembling those of a capacitor (high power, low energy storage) to ones that mimic a very high energy storage device (low power and high energy).

Probably, lithium-ion technology has more flexibility than any other battery system in the number of materials that can be used and the modifications that can be done at various stages of the materials processing (Fig. 3.1). Cathode materials can be a variety of oxides or even phosphates, anodes can range from carbon (most common) to metals or oxides, electrolytes can range from a variety of organic carbonate liquids with a myriad of potential additives to polymeric/gel type or solid and finally separators can range from plastics to ceramics. Considering the number of different additives to both anode and cathode and the fact that the material properties for each of these components, such as particle size, tap density, etc., are also variable, it is no surprise that an optimization around any specific application can take many years to complete. The task of optimization can be so complex that some companies tend to prefer vertical integration of their business model to include a material's business with a cell manufacturing business, since optimization of material properties can play such a large role in optimization of cell performance.

With the scope of optimization now explained, the intent of this chapter will be to drill down into specific examples how electrodes can be optimized to enhance a specific attribute of lithium-ion technology. Again, to give some understanding to the breadth of this issue of optimization, the following are some of the potential properties that might need to be optimized for a given application: charge and/or

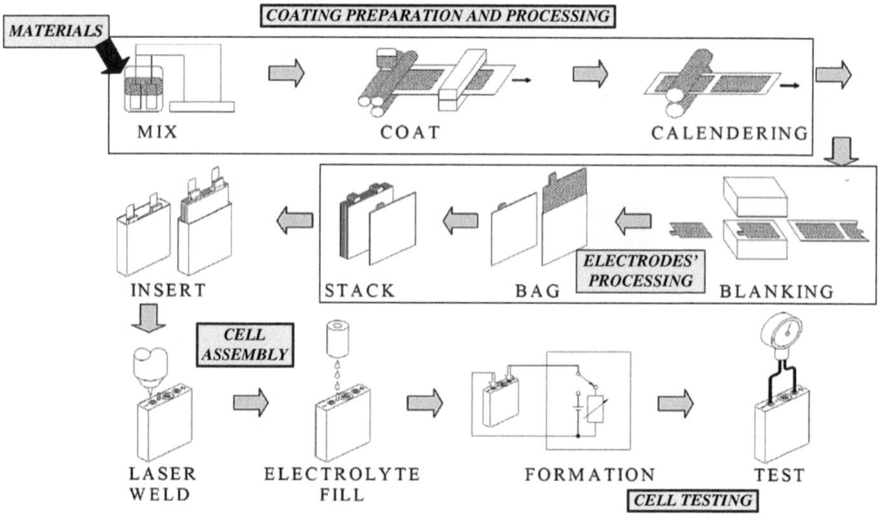

Fig. 3.1 Top-level process flow diagram for Li-ion cell assembly

discharge current rate (A), specific energy (Wh kg^{-1}) energy density (Wh L^{-1}), power (W), specific power (W kg^{-1}), power density (W L^{-1}), cycle life (usually defined as the number of cycles to 80 % of original capacity), temperature performance (performance at given low or high temperature), safety (nail penetration, shorting, overcharge, or incineration), and cost ($/any of the properties given above). To optimize the system around any of the properties given above, electrode formulations and cells design must be chosen accordingly.

3.2 Electrode Formulations and the Concepts of Weight Loading and Porosity

It was once believed that lithium-ion was the Holy Grail of battery technology that would solve all portable power needs for every possible application. Many soon realized that, like many complex systems, the enhancement of certain performance characteristics resulted in the detraction of other performance characteristics. High energy versus high power is the most common example of this dichotomy. To achieve high power, cell manufacturers may produce very thin electrodes that greatly increase the effective surface area of the electrodes thus reducing the effective current density (mA cm^{-2}) needed to achieve very high power output from the cell. The downside to this approach is the large quantity of electrochemically inactive current collector (Al for cathodes, Cu for anodes) that drives down the specific energy (Wh kg^{-1}) and energy density (Wh L^{-1}) of the cell. By contrast, another manufacturer could produce electrodes with very thick, very dense coatings, which maximizes the contents of the active materials inside the cell, resulting in a high energy density cell. However, such cell may have a large capacity yet only be able to utilize about 70 % of its capacity at the C rate (Fig. 3.2). Describing electrodes in terms of their weight loading (Text Box 3.1) provides a convenient way to quantify and compare rate capabilities of a given cathode and/or anode within a lithium-ion cell.

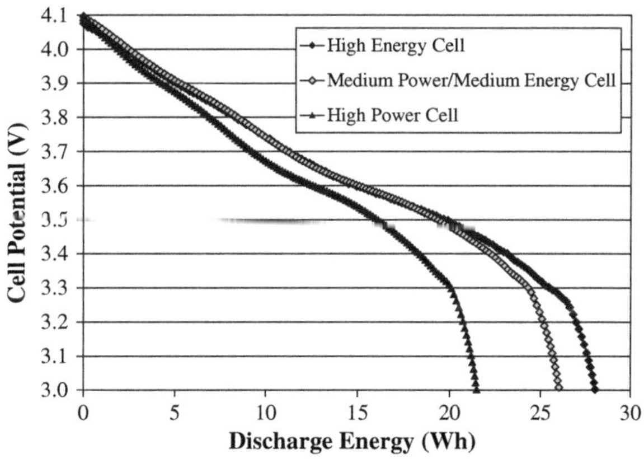

Fig. 3.2 Weight loading impact on discharge energy of a Li-ion cell

Text Box 3.1 Calculating electrode's weight loading

Weight loading is a measure of the mass of all the materials coated onto the metal foil substrate expressed in units of mg cm^{-2}. Both the measurement and the calculation of weight loading are very simple. A sample section of coating is stamped, punched out, or cut, having a known area in cm^2. The sample is then weighed. All of the coated material from one side of the coating is then removed (e.g., by scraping) until the metal foil substrate is bare. The single-sided sample is then reweighed. The mass of the coating is determined by the difference in the two weight measurements. The mass of the coating is divided by the area of the sample to get the weight loading of the first side of the coating (mg cm^{-2}). The entire process is repeated for the second coating side to obtain the corresponding weight loading value.

Porosity is another physical characteristic of an electrode that affects the electrode's power capabilities within the lithium-ion cell. During a slurry coating process (Fig. 3.1, the second step of the coating preparation and processing), the solvent used to dissolve the binder (the standard solvent for PVDF binders is *n*-methyl-2-pyrrolidone, NMP) is evaporated with heat and/or vacuum. The resulting dried coating has void spaces where the solvent was driven away from the coating, which constitute the electrode's porosity (Text Box 3.2). This structure often lacks the sufficient particle-to-particle contact between the active material and the conductive diluents, required to have high intra-electrode conductivity. By calendering the electrodes, the particle-to-particle contact is increased and porosity is decreased. Some porosity is required in order for electrolyte to transport lithium-ions to and from active materials for intercalation. By calendering to a specific porosity, one can ensure that an electrode will have high enough intra-electrode conductivity for electron transport as well as enough void space to allow electrolyte to transport lithium-ions to the active materials for intercalation.

Text Box 3.2 Calculating electrode porosity

Electrode's porosity is a measure of the theoretical void space of the materials coated onto the metal foil substrate, expressed as percentage of the total coated volume of (dried) materials. First, the thickness of a coated foil is measured in a variety of locations along the length and across the width of the coating. Using an average of these thickness measurements, the *total volume of the sample* is calculated. The *measured coating volume* is obtained on subtraction of the volume of the substrate (metal foil) from the total volume of the sample.

Next, the *theoretical coating volume* is calculated. First the composite density of the coated materials is obtained. The true density of each material, often referred to as particle density, is multiplied by the fraction of the material within the coating. The sum of these values is the *composite density of the coating*. Note: be sure not to use the tap or bulk densities of any material for these calculations. Finally, the previously calculated weight loading (Text Box 3.1) of the coating is multiplied by the area of the sample to obtain the *actual mass* of the coated material. The actual mass of the coated material is then divided by the composite density of the coating to obtain the theoretical coating volume.

The coating porosity equals the *measured coating volume* minus the *theoretical coating volume*; the subtraction result divided by the *measured coating volume*. The final result (porosity fraction of the coating) is often expressed as percentage of the measured coating volume.

The effect of weight loading on the discharge energy of a lithium-ion cell is illustrated in Fig. 3.2, where three lithium-ion cells of different designs are shown during their first discharge: a cell designed to deliver high energy, another cell that was designed for medium energy and medium power outputs and the cell that was built for high-power operation. All three cells in Fig. 3.2 were designated as INCP26/53/92 (per International Electrotechnical Commission, IEC [1]).

Text Box 3.3 Lithium-ion cell designations per IEC

INCP26/53/92 designation of the lithium-ion cell means that the cell has intercalation chemistry (I), includes a nickel cobalt-based cathode (NC) and is contained in the prismatic case (P) that is 26 mm thick × 53 mm wide × 92 mm tall.

All three cells utilized similar electrode materials (both active and inactive). All cells shown in Fig. 3.2 had $LiNi_{0.8}Co_{0.2}O_2$ (Chap. 1, Table 1.4) as the cathode active material and meso-carbon microbeads [2] (MCMB) as the anode active material. The high energy cell utilized MCMB 10–28 (Chap. 1, Table 1.7) while the other two cells utilized MCMB 6–28 (Chap. 1, Table 1.7). The binders were the same polymers from the same source in all cells, and the percentage of binder in the cathode was equivalent for all three cells; the same is true of the anodes. The set of conductive diluents used in all cells was the same across all three cell designs and the diluents' proportions to the active materials were also consistent. All cathodes were coated onto the same specification of aluminum foil, and the anodes used the same copper foil. The difference was the weight loading of the electrodes, and more specifically, the cathodes. The high power cell had the lightest weight loading. By comparison, the high energy cell had a weight loading that was three times as heavy. The medium energy/medium power cell had a weight loading that was 175 % of the high power cell. All electrodes were calendered, i.e., pressed between a pair of rotating cylinders (Fig. 3.1, third step of the coating preparation and processing), to a specified thickness. The desired thickness of each electrode was ultimately based on the optimum porosity, obtained as a result of prior art.

All of the cathodes and anodes were calendered to similar porosity values. The resulting thicknesses were what determined the number of pairs that could be fit into the metal cell case. The high energy cell had the thickest electrodes and therefore, the fewest count of pairs, 51. The medium energy/medium power cell had 84 electrode pairs, and the high power cell had the thinnest electrodes and 118 pairs. For every pair in a cell there were electrochemically inactive additional cell component materials: aluminum foil, copper foil, and separator material. The more pairs of electrodes were contained in a cell, the higher the percentage of inactive materials was additionally included in the cell. This is precisely the reason that three different cells with identical volumes had three different discharge energies. The resulting energy densities of the three cells varied in direct relation to their discharge energy since the cells' volumes were constant. The high energy cell delivered 274 Wh L^{-1}. The medium energy/medium power cell delivered 255 Wh L^{-1}. The high power cell delivered 210 Wh L^{-1}. Interestingly, all three cells had similar masses. The medium energy/medium power cell was only 2 % heavier than the high energy cell and the high power cell was only 4 % heavier than the high energy cell. The specific energy values of the high energy cell down to the high power cell were 114, 104, and 84 Wh kg^{-1}, respectively.

3.3 Weight Loading and Porosity of Electrodes:Impact on DC Resistance of the Li-ion Cell

The previous section demonstrated that by modifying the weight loading of electrode coatings, the specific energy and energy density of a given cell type can vary to a great extent. Weight loading and porosity were introduced as key physical parameters of an electrode. In the following section, examples are provided on how cathode's weight loading and porosity are optimized for a given cell design to improve the power capability of the cell. More specifically, the values of effective direct current (DC) resistance for three lots of cells are compared. There is a strong relationship between the cell's measured DC resistance value and the predicted power capability. To this effect, the DC resistance measurement done before any prolonged or strenuous testing of the cell is an "early indicator" of the cell's future power capability performance. Moreover, DC resistance measurements done over the span of the cell's life allow for tracking of the cell's "health" status.

Text Box 3.4 Effective DC resistance

The terms "resistance" and "impedance" are sometimes (incorrectly) used interchangeably. For the purposes of this book, resistance is the numerical result of a DC measurement and "impedance" is a phenomenon that can be observed (and modeled) using the Electrochemical Impedance Spectroscopy (EIS), a method that uses alternating current (AC) or alternating voltage signal to perturb the electrical circuit and measures the response of the circuit. For more detailed information on EIS, the reader is directed to the Further Reading section of this chapter.

The effective DC resistance is calculated by the application of Ohm's Law to the cycling data of a cell:

Ohm's law: $V = I \cdot R$

where V represents voltage (V), I represents electrical current (A) and R is a resistance (ohms).

A cell is discharging at a specified current (I_1) when a brief (in most cases, lasting 1 s) higher current (I_2) is applied to the cell. The absolute value of the change in voltage of the cell divided by the absolute value of the change in current equals the effective DC resistance:

$$\frac{|V_2 - V_1|}{|I_2 - I_1|} = R$$

The calculated effective DC resistance is often expressed in milliohms (mΩ) because (in a well optimized and properly built prismatic cell) the measured voltage change usually occurs in a millivolt range. Note: the milliohm range of DC resistance does not ally to coin cells and most experimental pouch cells; these test beds usually have higher resistance readings than the commercially manufactured, high-end prismatic cells.

State of charge, defined in Text Box 3.5, is another important characteristic of lithium-ion cell, which impacts the measurements of DC resistance values.

Text Box 3.5 State of charge and rated capacity

State of charge (SOC) is a measure of the percentage of capacity remaining in a cell and is dependent on several factors. To determine the SOC of a cell, the rated (initial) capacity of the cell must first be defined. Then, the operating voltage range for the cell must be established. Cell's past operating conditions such as the past charge/discharge current rates, temperature, and the overall "history of cycling" impact the SOC of the cell. Usually, SOC is rated either at the beginning of life (BOL) or end of life (EOL).

Figure 3.3 shows the mean DC resistance of three lots of cells at various states of charge. All cells were designated as INCP26/72/112. The cells shown in Fig. 3.3 were all rated according to Lot 1. The voltage range was 4.1–3.0 V. The discharge current rate was 6 A. The temperature was 20 °C. The cells were rated

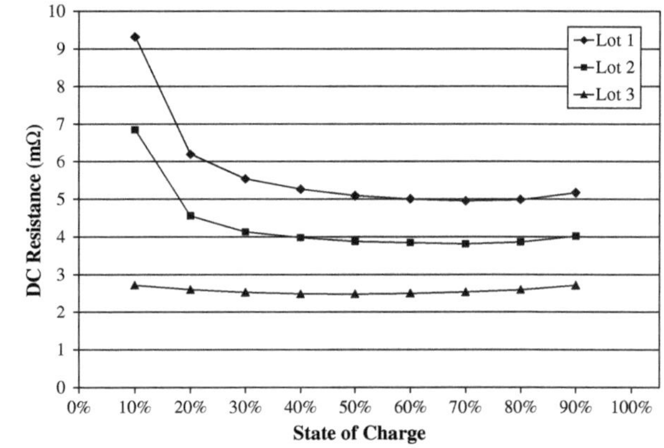

Fig. 3.3 Cathode porosity effect on DC resistance of Li-ion cells in various states of charge

at 12 Ah, BOL. All tested cells utilized the same electrode materials. All of the anodes used in cells were from a single coating lot. The cathodes all had the same formulation with respect to the percentages of the active material, binder and conductive diluents, and all used $LiNi_{0.8}Co_{0.2}O_2$ as the active material. The difference was the weight loading and porosity. Lot 1 had a weight loading of 9.61 mg cm^{-2} and a porosity of 48.9 %. Lot 2 had a weight loading of 9.61 mg cm^{-2} and a porosity of 42.6 %. Lot 3 had a weight loading of 10.85 mg cm^{-2} and a porosity of 35.2 %. Even with these variations in cathode, for the purposes of this experiment, all of the cells had the same number of pairs and therefore the same total electrode surface area. The reduction in porosity from Lot 1 to Lot 3 increased the particle-to-particle contact and therefore the intra-electrode conductivity, resulting in a lower effective DC resistance values across the span of various SOC points.

Figure 3.3 also demonstrates another important effect observed in lithium-ion cell cathodes. The effective DC resistance of each of the lots was higher at very high and very low states of charge. This effect was most pronounced in Lot 1. This is primarily a result of the change in conductivity of the mixed metal oxide cathode, $LiNi_{0.8}Co_{0.2}O_2$, as the level of lithiation changes.

DC resistance measurements, as mentioned before, can be done over the span of the cell's life and can assist in comparison of multiple cells' parameters and their impact on the final performance of the cell. Figure 3.4 gives an example of the DC resistance variations observed at 50 % SOC on the span of 4,500 cycles, where the experimental lots differed much more significantly than just in their weight loading and/or porosity values. In the experiment illustrated by plots in Fig. 3.4, various cathode and anode chemistries, and consequently, different binder contents were tested and compared. Single cathode/anode pair pouch cells were used in this experiment, with 4–6 cells built and tested per each experimental lot.

Lot 1 was based on the heritage electrodes' chemistries and formulations, based on $LiNi_{0.8}Co_{0.2}O_2$ cathode and an MCMB anode. Lot 5 was built with new cathode ($LiNi_{0.75}Co_{0.15}Al_{0.05}O_2$) and anode (synthetic graphite) chemistry and updated binder materials, with new binder contents roughly adjusted to accommodate the

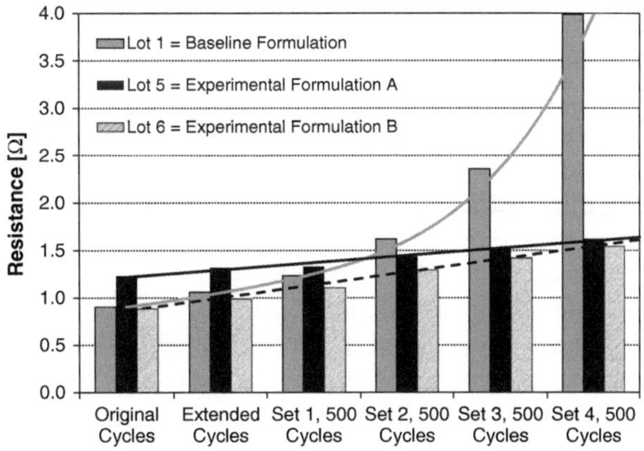

Fig. 3.4 DC resistance results at 50 % SOC measured in three experimental full pouch cell series

new binder's adhesion properties. Lot 6 was very similar to Lot 5, with the only difference being the type of anode material used, with smaller (~35 % less) average particle diameter of anode. The heritage chemistry performed well in the short term (up to about 500 cycles) and then the new lots have outperformed the traditional lot in terms of maintaining low values of DC resistance. When considering Lot 5 and Lot 6, an initial benefit of using smaller particles of anode in Lot 6 is apparent, with about 30 % lower DC resistance of Lot 6 measured at BOL. However, the results of longer term cycling (4,500 cycles) reveal that the slope of the linear DC resistance increase in Lot 5 is less inclined than the slope observed in DC resistance measurements for Lot 6, with both experimental lots reaching very close readings of DC resistance at 4,500 cycles. From the linear trend observed in both Lot 5 and Lot 6, a conclusion was drawn that Lot 5 would be a better candidate for long-term cycling (5,000 cycles and beyond). The difference in the performance of these two lots can be attributed to the fact that the same amount of binder used with anode particles that are about 35 % smaller (and have correspondingly larger surface area) is "consumed" while providing particle-to-particle bonds within the coating (cohesion) and less binder is available for the particle-to-foil contact (adhesion). Strong adhesion to the current collecting foil is known to be an important contributor to the long cycle life of a cell [3, 4].

3.4 Component Optimization, Example 1: Cathode and Binder Improvements

Most lithium-ion applications require the longest achievable cycle life. Depending on the conditions, what constitutes "long" is a relative term. As discussed in Chap. 2 of this book, it is possible to extend the cycle life of several lithium-ion cathode materials by restricting their cycling voltage range to the region of least entropy (least disorder induced by repeated intercalation of lithium ions) for a given cathode. Many satellites orbiting the Earth have lithium-ion batteries that have been designed to operate within the "minimal cathode entropy" voltage range. The battery may be required to utilize only 10 % of its capacity in the most life cycle-favoring section of its voltage range. In this case, the cycle life counted in tens of thousands of charge/discharge cycles takes precedence over maximizing the capacity/energy of the cell. After all, it is not easy to change the battery in a satellite orbiting our planet multiple times per day. An expectation of 40,000 cycles and 10 years of operation is not unusual. Other applications, due to weight or volume limitations, must use the smallest or lightest battery possible. Those applications may need to use 100 % of the battery's capacity for every cycle. In these instances, life expectancy of the battery is much shorter.

When selecting the operating voltage range for a battery, factors other than the longest cycle life or the maximum capacity sometimes influence the decision. There are cases, where the practical battery configuration determines the selection of the cell's operating voltage to a great extent (with the optimal voltage range still being kept in mind but not being the sole consideration anymore). For example, in a real-life system, if lithium-ion cells are connected six in series (6S) and the battery's charger has an upper limit of 25 V, then the mean cell voltage at the top of charge will be 4.166 V. In a 7S configuration, the upper voltage limit would be 3.57 V, resulting in a system with a very low energy density. In this particular system, the lower voltage limit of 2.85 V was selected as an attempt to maximize discharge energy for each cycle. Selected optimization results of two iterations of cells that were tested for this system are shown in Figs. 3.5 and 3.6.

Figures 3.5 and 3.6 show the discharge energy for two lots of lithium-ion cells. All cells were designated as INCP26/53/92. The cells were optimized for the greatest possible energy density. By comparison, an earlier generation of the high energy cell shown in Fig. 3.2, using the same size of the cell can, delivered less than 30 Wh during its first discharge. Obviously, the earlier generation of the high energy had different cell chemistry in terms of the cathode and anode active materials and different contents of binders and conductive diluents within the electrodes.

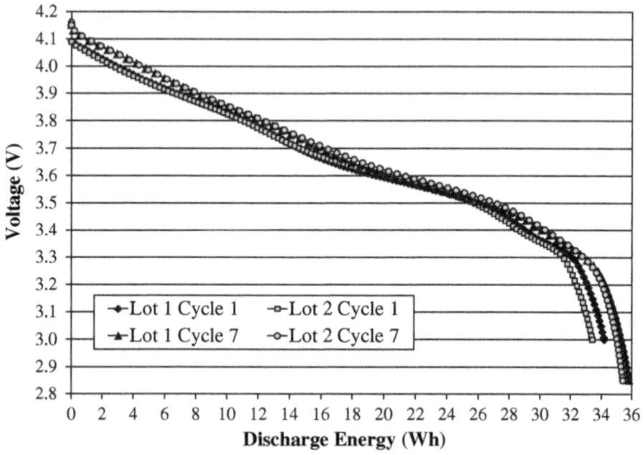

Fig. 3.5 The first and seventh discharge profile for two experimental cell lots

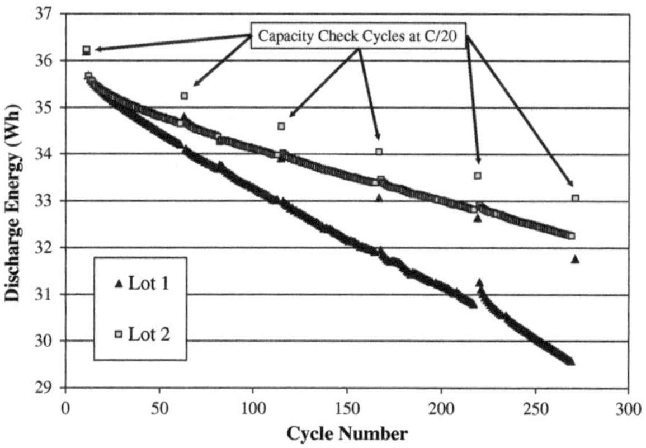

Fig. 3.6 The cycle life data for the two lots of cells

The two new lots (Lot 1 and Lot 2) of new generation high energy cells (Figs. 3.5 and 3.6) utilized identical anode coatings based. The cathodes were coated to the same weight loading, had the same conductive diluents in the same ratios, and the electrodes were calendered to the same thickness, which resulted in Lot 2 having a 1 % higher porosity. The major differences were the type and percentage of cathode material as well as the type and percentage of cathode binder. Lot 1 utilized $LiNi_{0.8}Co_{0.2}O_2$ as the cathode active material at 89.5 % with 6 % of PVDF binder. Lot 2 utilized $LiNi_{0.8}Co_{0.15}Al_{0.05}O_2$ as the cathode active material at 90.5 % with 5 % of a higher molecular weight PVDF binder. The reasons for the 1 % variation in cathode active and binder content were twofold. First, the aluminum in the Lot 2 cathode (5 mol % of the cathode) is electrochemically inactive, as mentioned in Chap. 1, Table 1.4. This has resulted in a slightly lower specific capacity for the Lot 2 material. Second, the increased molecular weight of the Lot 2 PVDF improved adhesion to the current collecting foil, allowing for less binder to be used.

Figure 3.5 shows the initial discharge cycle of a representative cell from each lot. The Lot 1 cell, as expected, had slightly higher discharge energy when charged to 4.1 V and discharged to 3.0 V at C/20 (cells rated at 8.5 Ah) and 25 °C. These cells were being optimized for the highest possible discharge energy while retaining the greatest possible cycle life. Figure 3.5 also shows the discharge voltage profile for the seventh cycle, in which the cells were charged to 4.166 V and discharged to 2.85 V at C/5 (cells were rated at 9 Ah in these conditions) and 25 °C. The reasons behind the selection of this particular voltage range were explained above by the battery configuration factors. This increased voltage range has increased the discharge energy of the Lot 1 cell by 5 % and the Lot 2 cell by 6 %.

Figure 3.6 shows the cycle life data for the two experimental lots of cells that were first compared in Fig. 3.5. The mean discharge energy of each cycle is displayed for each lot of four cells. The life cycling was done in a temperature chamber set to 35 °C. The cells were cycled in the same voltage range as that seen in Fig. 3.5, Cycle 7. The cells were rated at 9 Ah. They began cycle life testing with a C/20 capacity check cycle. They then cycled 50 times with a charge at C/7 and a discharge at C/5. Then, a DC Resistance test was done, similar to the testing shown in Fig. 3.3. The set of 50 cycles, followed by a DC resistance test and the capacity check was repeated five times in the testing shown in Fig. 3.6. Both lots of cells (Lot 1 and Lot 2) began the test delivering similar discharge energy. By the end of the first 50 cycles, the performance of the two lots was already beginning to diverge. At that point, Lot 2 delivered 1.3 % more energy under both high and low-rate cycling conditions. By 100 cycles, Lot 2 was delivering 2.9 % more energy at the higher rates and 2.0 % at the lower rates. This trend continued. After 250 cycles, Lot 2 was delivering 9.1 % more energy at the higher rates and 4.1 % at the lower rates.

Both lots were experiencing a loss of energy, but Lot 2 was losing almost the same energy at the higher rate, down 8.7 %, as it had at the lower rate, down 9.6 %. On the other hand, Lot 1 was experiencing a much higher loss of energy at the higher rate, down 17.1 %, than it had at the lower rate, down 12.2 %. This phenomenon is known as power fade, and represents cell's inability to discharge at high power. During the first capacity check cycle shown in Fig. 3.6, the cells were averaging 1.6 W. During the second cycle at C/5, the cells were averaging 6.6 W. Neither of these values would be considered high power in most lithium-ion systems but this was a system optimized for ultra high energy and power was the trade-off. Regardless of the cell design, investigation of power fade via DC Resistance measurements performed periodically on a cell is a great method to track and examine the changes in the cell's performance.

Figure 3.7 shows the mean DC Resistance for the two lots of cells during the cycle life testing including a linear fit of the data for each lot. The plot clearly shows the steady increase in the resistance of Lot 1. In fact, the slope of the linear fit of Lot 1 is 8.5 times greater than the slope for Lot 2. The cycle life testing of these cells was aggressive, as evidenced by two major factors. First, the high upper voltage limit (4.166 V) and wide voltage range (lower limit of 2.85 V) brought the active cathode materials through voltage sections with higher entropy. Second, the testing mode was continuous cycling in a chamber at 35 °C. The actual, internal temperature of the cells was bound to be even higher as the cells accumulated heat under the continuous cycling conditions. The improvements in the performance can be attributed to using a cathode material with greater stability at higher

Fig. 3.7 Mean DC resistance for the two experimental lots, measured in various stages of cells' life

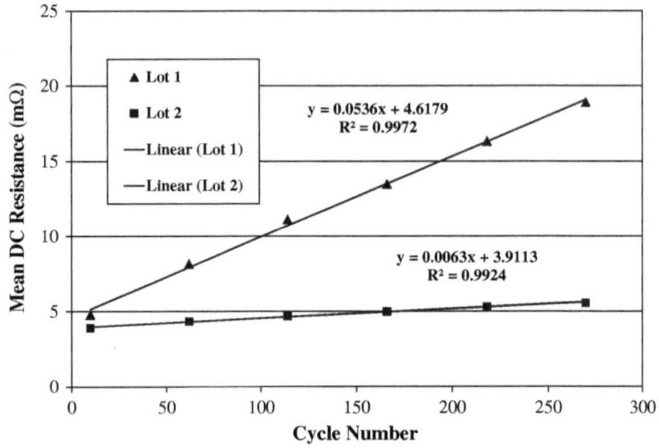

voltages for Lot 2 ($LiNi_{0.8}Co_{0.15}Al_{0.05}O_2$) and to the use of the PVDF cathode binder with higher molecular weight than the "standard" binder. The high molecular weight binder is less soluble in the organic carbonate electrolyte solutions at elevated temperatures and thus contributes to the extension of the cell's life.

3.5 Component Optimization, Example 2: Conductive Diluents in the Anode

The issues that some cathode materials have with low conductivity associated with certain levels of lithiation have been demonstrated. Commercial anode materials, which are mainly graphitic carbons, do not have the same problems with conductivity variations as cathodes. However, there is still a need for conductive diluents in the anode.

Figure 3.8 through 3.10 show data from three lots of cells with varying levels of conductive diluent in the anode. All cells were designated as INCP26/72/112. All cells utilized the same electrode materials. All of the cathodes were from a single coating lot using $LiNi_{0.8}Co_{0.2}O_2$ as the active material. All the anodes also utilized the same materials: MCMB 6–28 as the active material, the same grade of carbon black, and the same binder. The anode formulation for Lot 1 was 88 % active material, 9 % binder, and 3 % carbon black. The anode formulation for Lot 2 was 88.5 % active, 9 % binder, and 2.5 % carbon black. The anode formulation for Lot 3 was 90 % active, 9 % binder, and 1 % carbon black. The anodes were all calendered to the same thickness to ensure that the same number of pairs ended up in each cell. This resulted in slight variations in porosity due to the variation in density of MCMB 6–28 and carbon black. Lots 1, 2 and 3 were 35.5, 36 and 37.3 % porous, respectively.

Figure 3.8 shows the first cycle coulombic efficiency (discharge capacity divided by the charge capacity) of each of the three lots. Lithium-ion cells start their life having about 90 % reversible capacity and eventually achieve almost 100 % coulombic efficiency under repetitive cycling. The potential for very long cycle life for lithium-ion technology is a result of the very high coulombic and energy efficiencies. The first few cycles in the life of any lithium-ion cell are often referred to as formation. This is the time when the cell is irreversibly forming its solid electrolyte interface (SEI) layers on the surface of anode. Chapter 1 touched on the fact that many of the anode materials used in lithium-ion electrodes are unstable with respect to the components of the organic carbonates-based electrolyte and form protective layers originating from the decomposition of electrolyte materials: salts, solvents, and additives. In practice, all electrode materials form some kind of SEI layer that results in the consumption of electrolyte and the process is most pronounced at the beginning of life but can occur throughout the entire life of the cell. Figure 3.8 is a column chart displaying the first cycle coulombic efficiency of all of the cells with the mean value for each lot as a line overlaying the cell data. Lot 1, with the highest percentage of carbon black and the lowest percentage of MCMB,

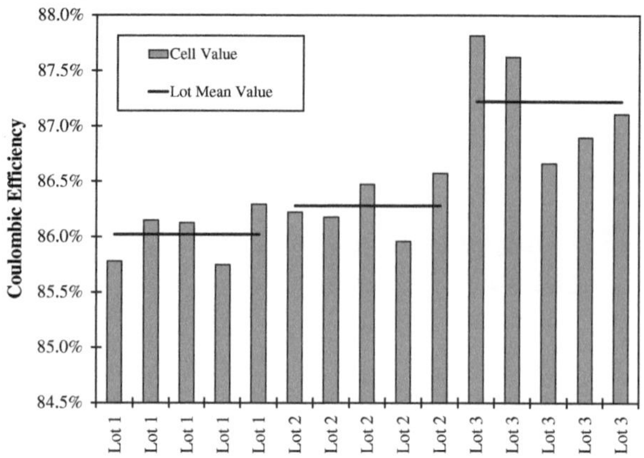

Fig. 3.8 Coulombic efficiency comparison in three developmental cell lots of the INCP26/72/112 model

had the lowest mean coulombic efficiency of 86.0 %. Lot 2 had 0.5 % less carbon black, 0.5 % more MCMB, and a mean coulombic efficiency of 86.3 %. Lot 3 had the lowest percentage of carbon black, the highest percentage of MCMB, and the highest mean coulombic efficiency of 87.2 %.

Given that the cathodes were identical in all three lots of cells, the large variation in first cycle coulombic efficiency can be explained by comparing the total surface area of the anodes. In Chap. 1, Table 1.7 it was noted that MCMB 6–28 has a surface area of 4 $m^2\ g^{-1}$. In Sect. 1.4, the surface area of carbon black was stated to be >50 $m^2\ g^{-1}$. For this experiment, the carbon black surface area was about 60 $m^2\ g^{-1}$. Therefore, Lot 2 had 5.3 % less anode surface area than Lot 1. Lot 3 had 21 % less surface area than Lot 1. The importance of total surface area of anode becomes clear when it is noted that SEI formation is mostly a surface reaction between anode and electrolyte. Variations in the first cycle coulombic efficiency are so important because the inefficiency during formation is irreversible.

After completing a number of formation cycles at 25 °C, all of the cells performed various cycles at rates up to C/2 (the cells were rated at 12 Ah, EOL), at temperatures varying from 0 to 20 °C. Figure 3.9 is a column chart displaying the discharge capacity of the ninth cycle, at C/2 and 20 °C, for all of the cells with the mean value for each lot as a line overlaying the cell data. Lot 1, having the lowest first cycle coulombic efficiency, had the lowest mean discharge capacity. Lot 2 had a mean discharge capacity 55 mAh higher than Lot 1. Lot 3, having the highest first cycle coulombic efficiency, had a mean discharge capacity 125 mAh higher than Lot 1. The results from this experiment presented thus far could be interpreted as follows: the less electrochemically inactive material (e.g., carbon black) is present in the anode, the higher the efficiency and, ultimately, the higher the capacity. And although that is true, there are other factors yet to be considered.

The application for the cells tested in this experiment required moderate discharge rates at low temperature. Specifically, the cells were discharged at 2 °C and −20 °C. Figure 3.10 is a plot of voltage versus time for a cell from Lot 1 and a

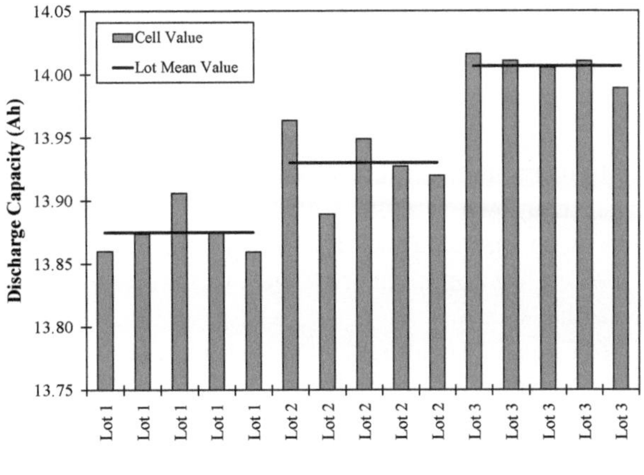

Fig. 3.9 Discharge capacity comparison in three developmental cell lots of the INCP26/72/112 model

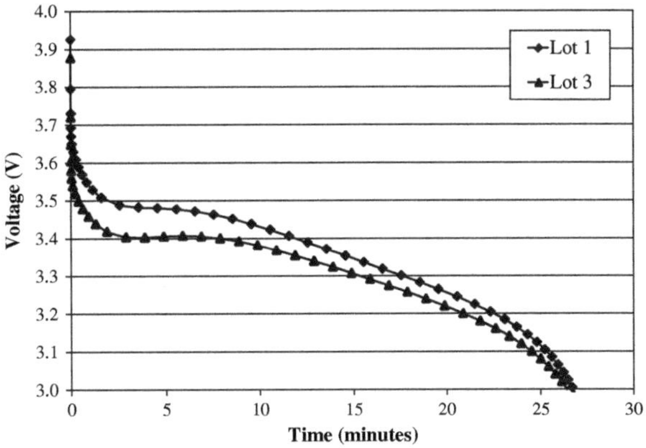

Fig. 3.10 Discharge voltage profiles developmental for two cells, each cell representative of their lot of origin; both cells were of the INCP26/72/112 model

cell from Lot 3. Although both cells delivered similar capacity, Lot 1 delivered slightly more: 10.7 Ah versus 10.6 Ah for Lot 3. More importantly, the Lot 1 cell experienced much less polarization and maintained higher voltage throughout the entire discharge, which resulted in the Lot 1 cell delivering higher discharge energy. This was exemplified by the cells' behavior during the initial discharge: four minutes into the discharge, the Lot 3 cell was at 3.40 V while the Lot 1 cell was still 80 mV higher at 3.48 V. The Lot 1 cell then went on to deliver 35.8 Wh versus 34.9 Wh from the Lot 3 cell.

This is a good example of another performance trade-off in lithium-ion technology. Reducing the amount of conductive diluents in electrodes improves the first cycle coulombic efficiency, but this occurs at the expense of higher rate capability and low-temperature performance. These effects are more apparent in the anode, in which the active material maintains relatively high, constant conductivity. In the cathode, where the active material has a comparatively lower conductivity that can vary with state of charge, the need for conductive diluents is much greater.

3.6 Systemic Optimization Example: Thermal Stability and Low-Temperature Performance

A wide operating temperature range is a goal for many lithium-ion technologies. Optimizations for low-temperature performance or for greater thermal stability pose their own sets of obstacles and potential solutions. An even more onerous task is to optimize electrodes to retain good low-temperature performance after repeated high temperature exposure. Every aspect of electrode construction must be addressed.

Figures 3.11 and 3.12 show the attempts to elucidate the problem. Figure 3.11 shows cycle life data for a baseline cell lot; all cells were designated as INCP26/53/92 and rated at 2 Ah. The entire volume of the cell case was not filled with electrodes, which would normally be the case with other prismatic cells. Instead, the cell in question was constructed with a specified number of electrode pairs sandwiched between a set of polymer shims. There were two very practical reasons for constructing the cells in this manner. First, the cycling equipment available to cycle the dozens of cells for this experiment was limited to charge and

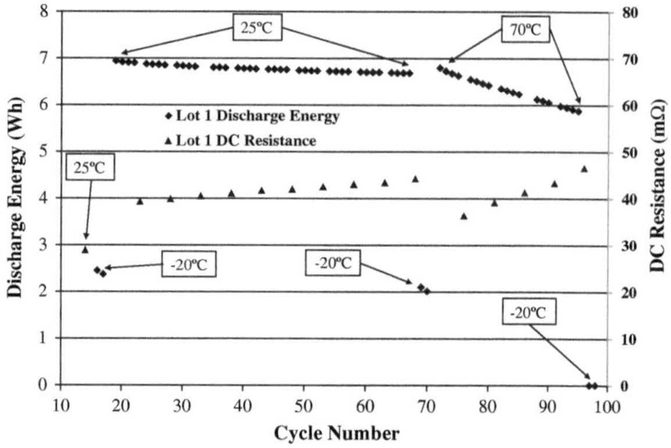

Fig. 3.11 Discharge energy and DC resistance plots for the baseline INCP26/53/92 cells at various temperatures

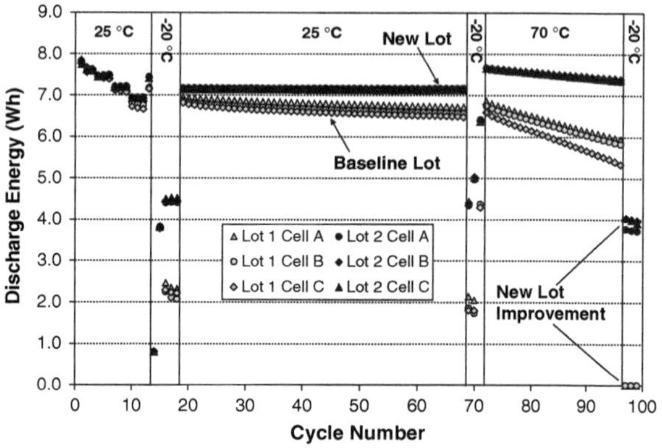

Fig. 3.12 Discharge energy comparison of the baseline and the "new lot" for INCP26/53/92 cells at various temperatures (modified after Gulbinska et al. [5])

discharge currents of 5 A. Second, the plan was to test the cell at temperatures as high as 85 °C. The use of an aluminized Mylar pouch cell was rejected out of concern that the cells would leak electrolyte at high temperature.

At elevated temperatures cathode and anode active materials experience an increased growth of resistive layers on the surface of their particles. The bulk of the material remains unchanged and continues to intercalate lithium-ions [6]. In the anode, the SEI layer thickens over time. While the cells remain at elevated temperature, the increasing resistance may not be easily observed. Once the cell is brought back to low temperature, the effects of the increased resistance are much more obvious and can result in a cell's failure to perform discharges at even modest rates without a severe reduction in delivered capacity. Binders that are not normally soluble in typical lithium-ion electrolytes could swell, deform, or even dissolve at elevated temperatures. The resulting damage to the electrodes ultimately leads to an increase in cell's resistance. Increased resistance is detrimental to high rate performance of the cell.

One of the goals of the optimization experiment shown in Figs. 3.11 and 3.12 was to greatly improve lithium-ion cell performance after exposure to very high operating temperatures (up to 85 °C) without losing the cells' ability to perform at −20 °C. These ambitious cell performance requirements called for comprehensive improvements in cathode and anode chemistry, electrode formulations, electrolyte additives, etc. The tested cathode candidates included lithiated mixed metal oxide ($LiNi_{0.8}Co_{0.2-x}M_xO_2$, where M = Al or Mg, with $0 \geq x \geq 0.15$) materials, whereas the anode enhancements focused on implementation of materials that were less sensitive to SEI layer side-reactions occurring at elevated temperatures and had more controlled particle size distribution (PSD). In addition to the active materials improvements, electrolyte enhancements were also extensively tested.

To date, the principal reason for using electrolyte additives in lithium-ion cells has been stabilizing the anode's SEI, which is critical to cyclability of Li-ion cells, because at elevated temperatures SEI deteriorates due to the reaction with $LiPF_6$-containing electrolytes [7–10] The loss of capacity and power from a cell stored at elevated temperature has been attributed to the presence of thermal decomposition products of the electrolyte in the anode SEI [10]. On the other hand, additives that inhibit the thermal decomposition of electrolytes with $LiPF_6$ were expected to improve the capacity retention and decrease resistance in thermally abused cells. Thus, the inhibition of the thermal reactions between the anode SEI and $LiPF_6$-containing electrolytes would lead to improved high-temperature performance of the tested cells. Since different families of additives have varied mechanisms of high-temperature performance enhancement, combinations of the two (or more) different additives were used in an anticipation of their cooperative effect.

The end result of the above efforts was a cell design that demonstrated very little degradation with extensive cycling up to 70 °C, while retaining performance at −20 °C that is comparable to the start of life performance of the baseline chemistry. It is worth noting that all experimental lots were comparable on the initial 25 °C test and that in the −20 °C test results the differences between experimental lots were more pronounced than at ambient temperature.

3.7 Separators Assessment and Optimization

The importance of lowering the internal resistance of the cell, especially when high power output is required, was touched upon earlier in this chapter. In fact, optimizing the conductive diluents and binder content within the electrode often boils down to striking the balance between developing the least resistive system that also has reasonable energy density. The main contributions to the cell's apparent (observed when measured) resistance are discussed in the Text Box 3.6. The topic of resistance is mentioned in the separator assessment and optimization section because very often the optimization of a cell for a given application requires the selection of the least-resistive, most permeable (to Li-ions) separator that also has a good mechanical stability and maintains a physical barriers between anode and the cathode.

Text Box 3.6 The main contributors to cell's resistance

The practical application of Ohm's law (Text box 3.4) to DC resistance measurements on the cell was also previously discussed. In reality, the IR voltage losses (also referred to as ohmic losses) constitute only a part of the total voltage loss experienced by a lithium-ion cell when the electrical load is applied. Mathematically, the following, simplified formula applies:

$$V = E_{\circ} - \left[(\eta_{ct})_a + (\eta_c)_a\right] - \left[(\eta_{ct})_c + (\eta_c)_c\right] - IR_I = IR \qquad (3.1)$$

where V is the cell's voltage that is lower than the standard cell potential $E_{\circ}$ or open circuit voltage (OCV) of the cell due to potential drops (polarization) caused by several factors, including:

1. Charge transfer overvoltage (also known as activation polarization) at the anode $(\eta_{ct})_a$ and cathode $(\eta_{ct})_c$
2. Lithium-ion concentration polarization at the anode $(\eta_c)_a$ and cathode $(\eta_c)_c$

Other terms used in Eq. 3.1 are the cell operating current I and the internal resistance of the cell R_I and the apparent (measured) cell resistance R. The formula is simplified in terms of only two kinds of polarization being included (the charge transfer polarization and the concentration polarization). The discussion of other types of polarization that also contribute to the cell's resistance is beyond the scope of this book and can be found in the sources quoted in the Further Reading section of this chapter.

The separator suppliers often provide their products with Gurley numbers which characterize air permeability (Chap 1 Table 1.10) of the films. The battery industry, however, is interested in the electric resistance of the separators wetted with the electrolyte solutions as discussed earlier. The electric resistance can be expressed in the units of specific area resistance (in Ω cm^2), or via MacMullin number which is the ratio of the resistance of the separator filled with electrolyte divided by the resistance of the electrolyte alone. Although, generally, the thinner

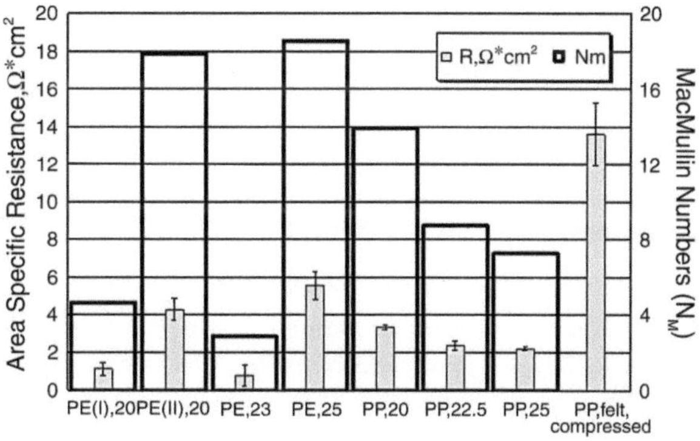

Fig. 3.13 Separators' resistance at room temperature. PP—polypropylene, PE—polyethylene; the numbers represent the thickness measured in μm (*courtesy of* Yardney Technical Products, Inc.)

separator is more permeable for diffusing ions, there is no simple direct correlation between the thickness and electric resistance (Fig. 3.13; data obtained at Yardney in 1 M $LiPF_6$ in EC/DMC/DEC (1/1/1 vol); a piece of the separator was compressed between two platinized platinum electrodes). MacMullin number is a temperature-independent parameter.

3.8 Electrolyte: Conductivity Optimization

As stated before in the concluding section of Chap. 1 of this book, the modeling efforts are expected to take increasing part in developing new materials for lithium-ion batteries. This is also true for the optimization and development efforts that focus on electrolyte solutions. Electrolyte properties can be modeled as an isolated entity but it is more useful for the battery manufacturer to consider the electrolyte as a component of the larger system (battery). For battery modeling purposes it is important to present the electrolyte solution conductivity (σ) as a function of concentration (C) and temperature (T). Thus, the software package Battery Design Studio® (CD-Adapco) offers several mathematical models. Two models were found most applicable by Yardney researchers: the Casteel-Amis (CA) model and the INL-Gering (INLG) model. The CA model is represented by the following formula:

$$\sigma = \sigma\mathrm{max}\left(\frac{C}{\mu}\right)^3 \exp\left[b(C-\mu)^2 - \frac{a(C-\mu)}{\mu}\right] \exp\left[-\frac{E_2}{R} - \left(\frac{1}{T} - \frac{1}{298.15}\right)\right] \tag{3.2}$$

where σ_{max} is the maximal conductivity at concentration C and μ, a, b as well as E_2 are the CA model fitting parameters.

The INLG model is represented by the following equation:

$$\sigma = \left[a \cdot \exp\left(\frac{-b}{T}\right) \cdot C^3 - d \cdot \exp\left(\frac{-e}{T}\right) \cdot C^2 + p \cdot \exp\left(\frac{-q}{T}\right) \cdot C\right] K \tag{3.3}$$

where $K = \frac{0.8}{\{1+\exp[5\cdot(C-2.1)]\}}$ and is expressed in (mS cm^{-1}) with fitting parameters a, b, d, p and q.

Obviously, regarding temperature, the CA model is nothing but the Arrhenius-type dependence [11]. The research group at Yardney has performed conductivity measurements of the solutions of more than ten different salts in a score of solvents and their blends. However, it has been proved by the Yardney results as well as by the available literature data that the Arrhenius behavior is at not typical for several lithium-ion batteries' electrolyte solutions. A few examples of the non-Arrhenius behavior are presented in Fig. 3.14.

Instead, the Vogel-Tammann-Fulcher (VTF) relation [12–14] (Eq. 3.3) has been found adequately describing the temperature behavior of the conductivity within the temperature range from −50 to 80 °C, at concentrations from millimol level to 2 mol per liter (Fig. 3.15) [15].

$$\sigma = AT^{1/2} \exp\left(-\frac{B}{T - T_0}\right) \tag{3.3}$$

where A, B and T_0 are constants. Although the exponential term in VTF equation looks similar to that in the Arrhenius equation, the B term does not have the same

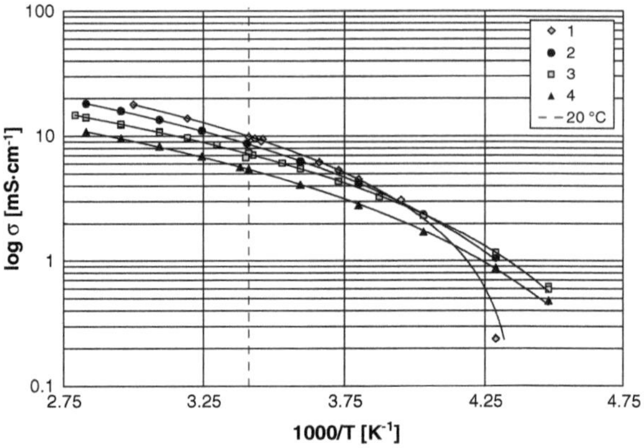

Fig. 3.14 Temperature dependence of the conductivity in Arrhenius coordinate set of the solutions: 1–1 M $LiPF_6$ in PC/DMC (1/1 mas); 2–1 M $LiPF_6$ in EC/DMC/DEC (1/1/1 vol); 3–1 M $LiPF_6$ in EC/DMC/DEC/EMC (1/1/1/2 mas); 4–0.5 M $LiPF_6$ + 0.5 M $LiBF_4$ in EC/EMC (1/3 mas)

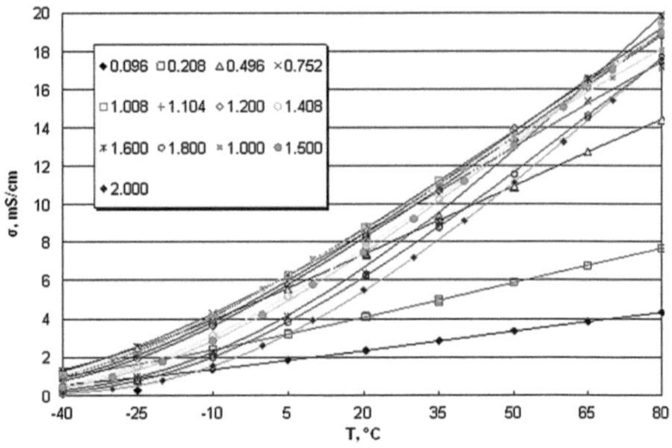

Fig. 3.15 Conductivity of $LiPF_6$ in EC/DMC/DEC (1/1/1 vol) at various concentrations, mol L^{-1}; points—experimental data; lines—VTF approximations

meaning of activation energy. B is related to the activation energy for the creation of critical free volume for ion transport. T_0 is the temperature at which the transport function ceases to exist or the solvent structural relaxation becomes zero. T_0 may be considered as the glass transition temperature for electrolyte solution or a temperature near it. In order to determine the parameters of the VTF equation, a nonlinear least-squares method must be employed [16].

The VTF parameters depend on composition of the solution. If the electrolyte concentration is the only composition variable the parameters can be represented as functions of concentration $A = A(C)$, $B = B(C)$ and $T_0 = T_0(C)$. The explicit forms of these functions can be revealed after obtaining the parameters' values at different concentrations. Thus, for the curves from Fig. 3.15, the correlation equations of the form: $A = AC^{1/2}, B = p + qC^{1/2} + rC$and $T_0 = u + vC^{1/2}$ lead to satisfactory results. Below, the VTF equation is given in the generalized form:

$$\sigma = aC^{1/2}T^{1/2}\exp\left(-\frac{p + qC^{1/2} + rC}{T - u - vC^{1/2}}\right) \tag{3.4}$$

The six equation (Eq. 3.4) parameters (a, p, q, r, u and v) need to be calculated. The result is presented in Fig. 3.16 where conductivities versus concentration at various temperatures are plotted. In the same chart, the approximations by INLG equation are also presented for comparison. Obviously, the same approach can be utilized if the composition is expressed by other characteristic than molar concentration. For example, the solvent's content can vary, including addition of

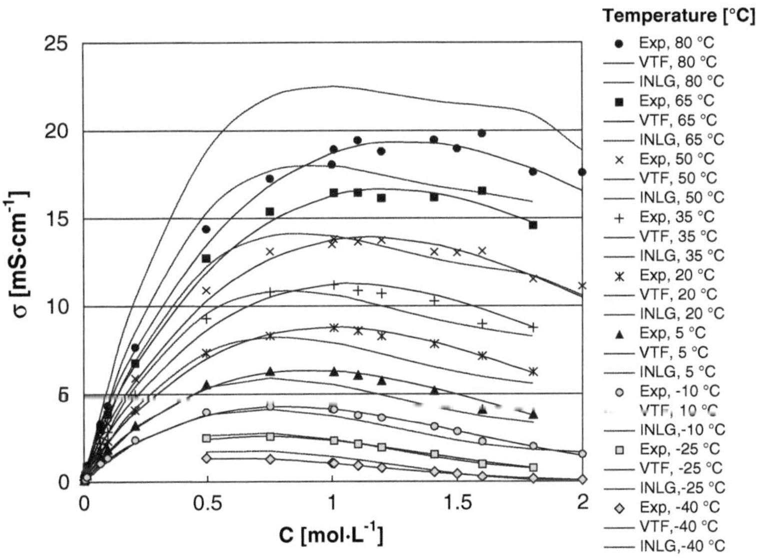

Fig. 3.16 Comparison of the generalized VTF and INLG approximations

flame-retarding or thermally-stabilizing components, or more than a single salt can be dissolved. The specific form of the correlation between the parameters and electrolyte composition would be different in each case.

3.9 Nanoparticulate Electrode Materials and Their Implementation and Optimization Challenges

Nanoparticulate electrode materials are considered by many as a viable solution to the limitations in the rate capability of Li-ion batteries [17–19]. The charge and discharge current rate limitations are mainly caused by slow solid-state diffusion of Li ions in the electrode materials [19, 20]. Amongst the potential advantages of nanoparticles are:

(1) Short Li-ion transport length due to small particle sizes. The Li-ion diffusion coefficient within a given solid, D_{solid}, determines the minimum particle size of the electrode materials required for their full utilization at the highest rates of charge and discharge [21]. $D_{solid} = l^2/t$, where l is the radius of an electrode active material particle and t is the time required for the Li ion to fully diffuse into it. When $D_{solid} = 10^{-9} - 10^{-10}$ cm^2/sec (e.g., in $LiCoO_2$, $LiNi_{0.33}Co_{0.33}Mn_{0.33}O_2$ and graphites), the average particles have to be in the 1–5 micron range or less in order to achieve full utilization in 60 s [22]. Therefore, great effort has been focused on the use of nanomaterials to improve the rate capabilities of lithium-ion electrodes.
(2) Fast surface reactions resulting from large electrode/electrolyte interface area.
(3) Good accommodation of structure strains imposed by electrochemical reactions [23].
(4) Possibility of operation in systems with low electronic conductivity due to short path lengths for electron transport.

The main reasons why nanomaterials show properties so different from those of the bulk are [24]:

(1) *Surface effects* ("trivial" size effects which are caused solely by the increased surface to volume ratio in nanoparticles). Atoms at surfaces have fewer neighbors than atoms in the bulk. Because of this lower coordination and unsatisfied bonds, surface atoms are less stabilized than bulk atoms. The surface-to-volume ratio scales with the inverse size, and therefore there are numerous properties which obey the same (linear) scaling law, such as the melting and other phase transition temperatures.
(2) *Quantum size effects* ("true" size effects, which involve changes of local materials properties). In metals and semiconductors (note: cathode materials for lithium-ion batteries can be considered semiconductors for practical purposes) the electronic wave functions of conduction electrons are delocalized over the entire particle and there is a continuous size-dependence of various materials' characteristics in the non-nano region. In nanomaterials, a discontinuous behavior of quantum size effects is observed, with the ionization potentials and electron

affinities for electrons involved in redox reactions no longer exhibiting a non-continuous size-dependence. This means that there is a reason to expect different reaction paths and reaction mechanisms in a nanoscale material from its bulk (micrometer-scale for the most Li-ion materials) counterpart.

Despite many potential advances in using nanoparticles in lithium-ion cells, there are three distinct obstacles associated with nanostructured electrodes that are very often quoted in the popular literature. First, the increased electrode/electrolyte interface area leads to significant undesirable electrode/electrolyte side reactions, safety concerns and poor calendar life; second, the lack of control over the synthesis process and high expense to fabricate the electrode hinder the progress toward large-scale production. Third, the use of nanoparticles leads to lower packing density in the electrodes, thus reducing the energy density of batteries [25]. From the lithium-ion cell manufacturing point of view, there is a forth, nontrivial obstacle to nanoparticles' implementation: imperfect compatibility of nanoparticulate materials with the existing battery making process. This obstacle is rarely listed in scientific articles as they often originate from the academic environment where there is limited awareness of the battery manufacturing process and the challenges associated with scaled-up materials handling and processing. One of the major reasons for this book is to promote the information exchange between academia and industry.

The common incompatibility of nanoparticles with the currently used battery making process manifests in the necessity for much more energetic mixing procedures than the ones normally used in battery making process (Fig. 3.1, coating preparation and processing—the "mix" step). Very often, a propeller-type mixer is sufficient for the "standard" battery materials, with the high-shear mixing or blending of powders being called for very occasionally and for limited duration. This is not the case with nanoparticles that often require the prolonged high-shear mixing of slurry (NMP plus active material, diluents and binder), which tends to significantly elevate the temperature of the slurry and break down the polymeric structure of PVDF binder. Obviously, auxiliary equipment can be used to alleviate the temperature rise, such as cooling jackets around the mixing vessel etc. but this adds another level of complexity (and cost) to the process and does not always completely solve the problem as the localized overheating could still occur.

Another incompatibility of nanoparticles with the commercial battery making process is caused by the excessive "soaking up" of the solvent/dispersant NMP by the large surface-area nanoparticles, which results in the NMP:solids weight ratio changing from about 50:50 to the ratio as high as 90:10. The high NMP content results in a very different rheology of slurry (e.g., very low viscosity) and causes excessive solvent waste generation (increased cost and environmental concerns). Very often, additional amounts of PVDF binder are also needed for sufficiently binding nanoparticles to the current collector due to their high surface area and this, to some extent, negates the previously described extensive optimization efforts on electrodes' formulation. Last but not least, the matter of the (yet) incompletely explored health hazards and the potentially insufficient protective equipment for handling nanoparticles is still unresolved. The need for the adequate protective equipment (respirators, gloves, hazmat suits, etc.) and rigorous ventilation procedures is clear when one considers the massive amounts of powders processed at the battery plants (tens to hundreds of kilograms of powders per coating). The above cautionary points are not meant to dissuade from pursuit of nanoparticles in batteries. There is certainly a great potential in these materials but the benefits should always be weighed against the disadvantages. Another point that can be made is the increasing need for the properly trained battery process engineers and the environmental health and safety specialists with the nanomaterials handling knowledge.

3.10 Conclusions

In this chapter, the process of optimizing the lithium-ion cells was discussed and specific examples of developing the specialized high-end lithium-ion cells were given. As long as there are new materials being introduced into the market for application in batteries, there will be the need for the new materials implementation and re-optimization of the cell designs. Performance and component modeling as well as the accelerated testing regimens will be continuously improved to suit the needs of battery industry for fast turnaround times, low cost and high accuracy of these methods. At some point, alternative coating methods may become more viable than the slurry-based coating process which could be replaced with solvent-less direct material deposition methods such as chemical vapor deposition (CVD) and others (pyrolysis-based methods, etc.). Currently, the cost and engineering challenges with the mass scale-up of these methods are the most prohibitive factors.

3.11 Further Reading

Electrochemical Impedance Spectroscopy (EIS)

(1) Lasia A (1999) Electrochemical impedance spectroscopy and its applications. In: Modern Aspects of Electrochemistry, Conway BE, Bockris J, White RE, (eds) Kluwer Academic/Plenum Publishers, New York

Thermodynamics of Batteries (Including Polarization Types and Origin)

(1) Bagotsky VS (2006) Fundamentals of Electrochemistry 2nd Edition. John Wiley & Sons, Inc. Hoboken, NJ

Modeling of Electrochemical Processes and Transport Phenomena

(1) M. Park et al. (2010) A review of conduction phenomena in Li-ion batteries. J Power Sources 195:7904-7929

General nanoparticles literature

(1) Sides CR et al. (2002) Nanoscale materials for lithium-ion batteries. MRS Bull 26:604-608. doi:10.1557/mrs2002.195
(2) Fichtner M (2011) Nanoconfinement effects in energy storage materials. Phys Chem Chem Phys 13:21186–21195. doi:10.1039/c1cp22547b
(3) Sides, CR et al. (2006) Nanoscale materials for lithium-ion batteries. MRS Bull 27:604-607. doi:10.1557/mrs2002.195
(4) Bazito FFC, Torresi RM (2006) Cathodes for Lithium Ion Batteries: The Benefits of Using Nanostructured Materials. J Braz Chem Soc 17:627-642

Specific examples of nanoparticulate electrode materials

(1) Li J, Tang Z, Zhang Z (2005) H-titanate nanotube: a novel lithium intercalation host with large capacity and high rate capability. Electrochem Commun 7:62-67. doi:10.1016/j.elecom
(2) Hosono E et al. (2006) Fabrication of nano/micro hierarchical Fe_2O_3/Ni micrometer-wire structure and characteristics for high rate Li rechargeable battery. J Electrochem Soc 153:A1273-A1278. doi:10.1149/1.2195887
(3) Yan L et al. (2012) Chemistry and physics of a single atomic layer: strategies and challenges for functionalization of graphene and graphene-based materials. Chem Soc Rev 41: 97-114. doi:10.1039/c1cs15193b

References

1. Reddy TB, Linden D, eds (2010) Linden's handbook of batteries 4th edition. McGraw-Hill, New York
2. Yamada Y et al (1974) Characteristics of meso-carbon microbeads separated from pitch. Carbon 12:307-319. doi:10.1016/0008-6223(74)90072-4
3. Jarvis CR et al (2001) Use of grafted PVdF-based polymers in lithium batteries. J Power Sources 97:664-666. doi:10.1016/S0378-7753(01)00696-6
4. Wu HC, Wu HC, Lee E, Wu NL (2010) High-temperature carbon-coated aluminum current collector for enhanced power performance of LiFePO4 electrode of Li-ion batteries. Electrochem Commun 12:488-491. doi:10.1016/j.elecom.2010.01.028
5. Gulbinska M et al (2011) Comprehensive improvements in Li-ion batteries for demanding applications. J Power Sources 196:2899-2904. doi:10.1016/j.jpowsour.2010.10.087
6. Dupre N et al (2010) Aging of the LiFePO4 positive electrode interface in electrolyte. J Power Sources 195:7415-7425. doi:10.1016/j.jpowsour.2010.05.042
7. Lee HH et al (2005) The function of vinylene carbonate as a thermal additive to electrolyte in lithium batteries. J Appl Electrochem 35:615-623. doi:10.1007/s10800-005-2700-x
8. Xu K et al (2005) LiBOB: Is it an alternative salt for lithium ion chemistry? J Power Sources 146:79-85. doi:10.1016/j.jpowsour.2005.03.153
9. Zhang SS (2006) An unique lithium salt for the improved electrolyte of Li-ion battery. Electrochem Commun 8:1423-1428. doi:10.1016/j.elecom.2006.06.016
10. Li W et al (2006) Lithium-ion batteries: Thermal reactions of electrolyte with the surface of metal oxide cathode particles. J Electrochem Soc 153:A1617-A1625. doi:10.1149/1.2210588
11. McNaught AD and Wilkinson A (1997) IUPAC Compendium of Chemical Terminology, 2nd edition (the "Gold Book"). Blackwell Scientific Publications, Oxford. http://goldbook.iupac.org. doi:10.1351/goldbook. Accessed May 2013

12. Gu GY, Laura R, Abraham KM (1999) Conductivity-temperature behavior of organic electrolytes Electrochem Solid St 2:486-489. doi:10.1149/1.1390879
13. Gu GY et al (2000) 2-Methoxyethyl (methyl) carbonate-based electrolytes for Li-ion batteries. Electrochim Acta 45:3127-3139. doi:10.1016/S0013-4686(00)00394-7
14. Blomgren GE (1999) Physical and chemical properties of non-aqueous solutions. In: Non-aqueous Electrochemistry. Aurbach D (ed) Marcel Dekker, New York - Basel
15. Ravdel B, Abraham KM, Gitzendanner RL, Marsh C (2003) Temperature dependence of the conductivity of lithium-ion battery electrolytes. In: Proceedings of the Symposium on Batteries and Supercapacitors. Nazri GA, Takeuchi E, Koetz R, Scrosati B (ed) The Electrochemical Society
16. Himmelblau DM (1972) Applied Nonlinear Programming, Mc Graw Hill, New York
17. Duan H et al (2008) Fabrication and characterization of Fe3O4-based Cu nanostructured electrode for Li-ion battery. J Power Sources 185:512–518. doi:10.1016/j.jpowsour.2008.06.07
18. Arico AS et al (2005) Nanostructured materials for advanced energy conversion and storage devices. Nature Mater 4:366-377. doi:10.1038/nmat1368
19. Sides CR et al (2002) Nanoscale materials for lithium-ion batteries MRS Bull 26:604-608. doi:10.1557/mrs2002.195
20. Malik R, Burch D, Bazant, M, Ceder G (2010) Particle size dependence of the ionic diffusivity. Nano Lett 10:4123-4127. doi:10.1021/nl1023595
21. Bruce PG (1997) Solid-state chemistry of lithium power sources. Chem Commun 19:1817-1824
22. Abraham KM (2009) The 2009 Fall ECS meeting, Vienna, Austria, Abstract No. 692
23. Szczech JR, Jin S (2011) Nanostructured silicon for high capacity lithium battery anodes. Energy Environ Sci 4:56-72. doi:10.1039/c0ee00281j
24. Roduner E (2006) Size matters: why nanomaterials are different. Chem Soc Rev 35:583–592. doi:10.1039/b502142c
25. Malik R, Burch D, Bazant, M, Ceder G (2010) Particle size dependence of the ionic diffusivity. Nano Lett 10:4123-4127. doi:10.1021/nl1023595

Chapter 4
Lithium-ion Cells for High-End Applications

Gregory J. Moore, Frank J. Puglia and Malgorzata K. Gulbinska

Abstract Engineering and building lithium-ion cells for high-end applications usually requires a very different approach than manufacturing commercial batteries. The high-end power sources engineering challenges and ways to address them are in most cases unique to a given application and often require "pushing the envelope." The lessons learned in such challenging conditions are often transferable to commercial applications. In this chapter, lithium-ion batteries for satellites, high energy, high power, and elevated temperature applications are discussed.

4.1 Lithium-ion Cells for High-End Applications: Introduction

Due to their high energy density and power capabilities, lithium-ion batteries are used for a variety of applications, including high-end applications that extend well beyond consumer electronics and hybrid vehicles where cost is of primary importance. High-end applications often serve military needs on the land, in the air, on and under the water; enable space travel and exploration and power satellites. There are numerous companies that make deep sea exploratory vehicles for civilian applications, e.g., searching for new pockets of oil or supporting the current oil production by petroleum companies. Other users are interested in deep sea vehicles for investigating sunken crafts. The current examples of robotic exploration vessel makers include iRobot Corporation, Liquid Robotics, Schilling Robotics, LLC, and Teledyne Technologies, Inc [1]. High-end applications also include such specialized niche markets as human implantable lithium-ion cells for powering a range of medical devices.

G.J. Moore (✉) · F.J. Puglia · M.K. Gulbinska
Yardney Technical Products, Inc., South County Trail, 2000, East Greenwich, RI 02818-1530, USA
e-mail: gmoore@yardney.com

M.K. Gulbinska (ed.), *Lithium-ion Battery Materials and Engineering*, Green Energy and Technology, DOI 10.1007/978-1-4471-6548-4_4

Table 4.1 Selected specifications of several Li-ion cells made at Yardney

		Discharge rate (A)		Discharge temp.	Energy density	Specific energy
Cell ID	Capacity, Ah[a]	Continuous[a]	Pulse[a, b]	°C	Wh L^{-1}	Wh kg^{-1}
NCP0.05-1	0.05	0.25	0.5	−20 to +60	175	60
NCP55-1	50	200	400	−40 to +70	265	112.5
NCP8-1	8	40	80	−40 to +60	320	130
NCP25-1	30	125	250	−20 to +50	335	145
NCP25-2	25	250	1,250	−40 to +60	250	105
NCP350-1	350	120	–	−20 to +60	417	212

[a]25 °C
[b]0.1 s

The driving requirements of the many types of high-end applications supported by lithium-ion technology vary substantially with regard to their need for power, energy, volume, and weight constraints. Other differentiating factors include the operation and storage temperature windows. The means of addressing the safety issues and the safety challenges themselves also differ between applications. Fortunately, with many types of cathode active materials, separators, and variations of electrolyte compositions with different solvents, salts, and additives, the capability window for lithium-ion technology is huge and can frequently address the most challenging requirements [2]. It is with this capability that a battery manufacturer working with a prime for high-end applications undertakes the responsibility to provide a battery system that will perform to meet or exceed the expectations. In this chapter, several high-end applications will be discussed using several examples to illustrate the dominating trends in the field of high-end lithium-ion technology. Specifications that set apart the high-end devices from their commercial counterparts will also be enumerated. With some products, the narrowing of the performance gap between the high-end and the commercial batteries should be noted. The discussions below are based on fielded products that show real solutions used in practice, as well as the developmental units available in limited quantities and, finally, the concepts that have not yet been implemented.

The extent to which various lithium-ion products can differ in many of their properties is illustrated in Table 4.1. Even the most basic property, such as capacity, spans from the 0.05 Ah cell for medical applications (middle ear implants) to the 350 Ah cell used in long-range propulsion for naval applications [3]. The difference in capacity is directly tied to the varying physical size of cells, as seen in Table 4.1 and in Fig. 4.1, where the largest shown cell has about seven hundred times the volume compared to the smallest implantable cell. The curved bottom face of both cells shown in Fig. 4.1 is coincidentally similar and spurs from the ergonomic design of both cells to best fit the available space (the submersible vessel or the human skull, respectively). One of the advantages of the true prismatic lithium-ion cells is their relatively conformal shape that can be relatively easily designed to snugly fit the confined spaces, with consideration given both to the

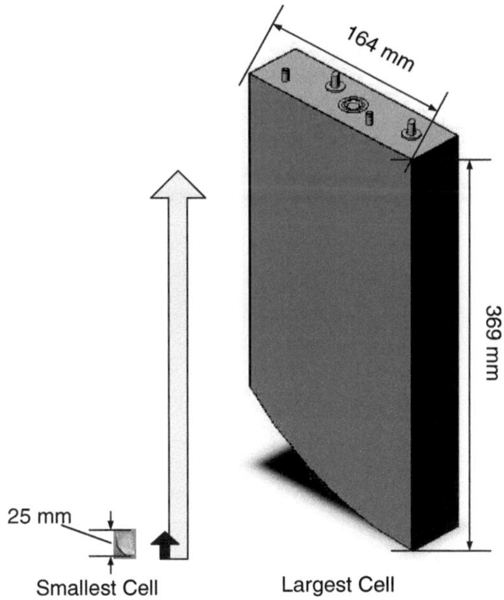

Fig. 4.1 An example of the wide span of dimensions of Li-ion cells (courtesy of Yardney Technical Products, Inc.)

physical fit and to the thermal conduction properties of the entire battery design. The thermal properties of a given battery design significantly impact the safety of the entire system and will be commented upon in Chap. 5 of this book, related to safety of lithium-ion batteries.

Specific energy (Wh kg^{-1}) and energy density (Wh L^{-1}) are the most commonly compared specifications of various cells (or batteries). When comparing the energy density or specific energy of various units, one ought to state whether the value refers to the cell level or the entire battery. In the case of a battery, the electronics and additional packaging as well as the potential auxiliary cooling accessories (pumps, tubing, cooling channels, etc.) would effectively lower both the specific energy and energy density compared to their cell level values. The graph shown in Fig. 4.2 illustrates how different application environments, ranging from implants to satellites, require different cell capabilities and packaging which causes major differences in their normalized energy, calculated per mass or per volume.

Charge and discharge current rates differ significantly depending on the lithium-ion battery application and resulting cell/battery design. Usually, the range of recommended charge current rates is restricted (often lower than C-rate), since high charge rates negatively impact the cycle life of a battery. At excessively high charge rates and/or at very low temperatures (e.g., subzero), there is a possibility of lithium metal plating, likely in the form of lithium dendrites, which constitutes a safety hazard. Of course, the exact C-rate and temperature conditions that bring the battery out of the "safe zone" vary between different cell designs and are outside of the manufacturer-defined operating parameters.

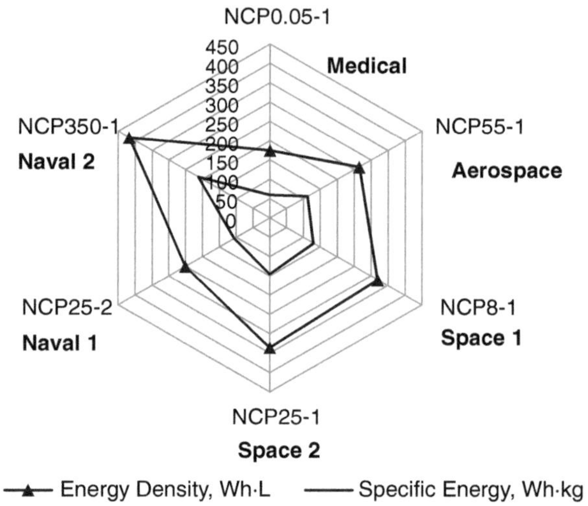

Fig. 4.2 High-end Li-ion cells energy comparison using a hexagonal chart with 50 unit increments (courtesy of Yardney Technical Products, Inc.)

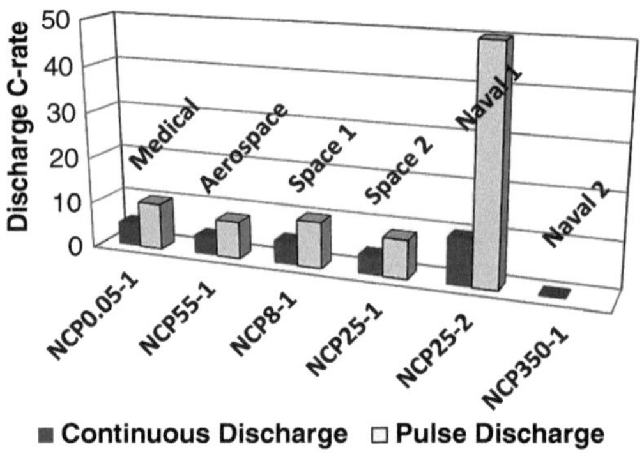

Fig. 4.3 High-end Li-ion cells discharge rate comparison (courtesy of Yardney Technical Products, Inc.)

Recommended discharge current rates are strongly related to the cell design (high rate or high energy) and to the desired performance parameters, such as end of discharge voltage, etc. Several examples of various Li-ion products, along with their continuous and pulse discharge characteristics, and their applications are shown in the graph in Fig. 4.3. As seen in the column chart, the general application environment does not necessarily translate to the particular set of discharge characteristics. For example, there is a vast difference between the two different naval applications shown in Fig. 4.3 plot. The difference spurs from the fact that the first application is a training torpedo that travels relatively short distances and calls for quick bursts of energy and can be retrieved and recharged after use. The second naval application is a manned vessel that travels reasonably long distances but at a steady pace and calls for the maximum energy density, while sacrificing the power capabilities. This was shown in Fig. 4.2, where the long-range application reaches the extreme limits in terms of the energy density and specific energy ratings. Properties of these different types of cells are listed in Table 4.2.

The bar graph presented in Fig. 4.4 illustrates the variations in temperature range of different lithium-ion high-end batteries. There are several determining factors

Table 4.2 Selected application requirements (courtesy of Yardney Technical Products, Inc.)

Cell ID	Defining application requirements
NCP0.05-1	Medical: compact, reliable, sterilized, good pulse performance
NCP55-1	Aerospace: medium energy, medium rate
NCP8-1	Space 1: high energy, medium rate, wide temperature range (mars exploration rover)
NCP25-1	Space 2: high energy, medium rate, medium temperature range (satellites, mars lander)
NCP25-2	Naval 1: short range, medium energy, high rate
NCP350-1	Naval 2: long range, very high energy, very low rate

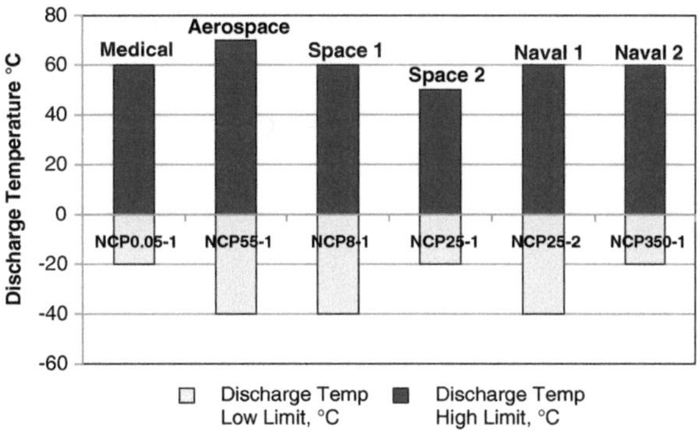

Fig. 4.4 High-end Li-ion cells temperature ranges comparison (courtesy of Yardney Technical Products, Inc.)

in the final recommendation for the battery's operating temperature; the few most important factors being: the electrolyte chemistry and additives (used either for high temperature or low temperature performance enhancements) and the potential thermal insulation and/or heaters built in the battery package. A high-performance insulation and/or heater can significantly narrow down the required temperature range, as it is in the case of one of the space applications ("Space 2") compared to much more demanding Aerospace application temperature range. The overall summary of various Yardney cells along with keywords describing their most important properties is presented in Table 4.3. In order to provide an industry-wide perspective, another comparison was done (Table 4.4) based on publicly available technical specifications contrasting the products from the leading lithium-ion battery manufacturers (commercial and high-end) with the Yardney products discussed previously.

4.1.1 Land

High-end land applications, which are mostly military, consist of land vehicles, man power, and weapons applications. Typical vehicle batteries consist of a long-time standard lead-acid, because of their low cycle life cost, which is based on the quantity of energy received from the battery over its full lifetime. Advantages

Table 4.3 Yardney's Li-ion cells compared to commercial prismatic cells

Cell model/maker	Capacity, Ah	Energy density, Wh L^{-1}	Specific energy, Wh kg^{-1}	Continuous discharge, C-rate	Temperature range[a], °C
High-end Li-ion prismatic cells range[b]/Yardney	0.05–350	175–417	60–212	0.3–200[c]	−40–70
Li-ion Gen IV prismatic/Hitachi (high power)[d] [4]	4.8	89	72	63	−30–50
Li-ion EV/PHEV prismatic/hitachi (high energy)	30	265	130	16	−30–50
Li-ion LSE190 prismatic cell /GS Yuasa (energy optimized) [5]	190	315	164	1	−10–35
Li-ion UF103450P cell/panasonic (energy optimized) [6]	1.9	397	180	2	−20–60
Li-ion ICP103450 prismatic/ samsung [7]	2.0	404	185	0.2	–

[a]Developmental, evaluation samples available
[b]Values are based upon seven different product lines; see Table 4.1 for the detailed breakdown of the group
[c]Developmental, prototypes under external evaluation
[d]In commercial cells, various temperature ranges are usually given: charging range, discharging range, and storage range

Table 4.4 Li-ion multi-kW modules—Ultralife (military land applications)

Part number	URB0001	URB0003
Capacity	100 Ah	140 Ah
Energy	2,560 Wh	3,710 Wh
Energy density	73 Wh kg^{-1}	103 Wh kg^{-1}
Weight	35 kg (77 lb)	~38.5 kg (85 lb)
Max. discharge current	100 A continuous	
Max. pulse discharge current	125 A for 2 s, 200 A for 500 μs	
Charging	Recommended charge rate is 12 A to 28.8 VDC within temperature range 0 to 45 °C	
Fast charging	20 A at 25 °C. Charge rate can be up to 100 A; however a charge rate over 20 A may reduce cycle life	
Voltage range	24.0–28.8 V	
Average voltage	25.6 V	
Cycle life	<3000 cycles at 1C to 80 % of initial capacity	
Operating temperature	−25 to 65 °C	
Storage temperature	−30 to 70 °C	
Self-discharge	<5 % per month	
Over voltage limit	29.2 V	
Under voltage limit	21.6 V	
Exterior/housing	ABS plastic	
Other characteristics	Wide operating temperature range	
	Forced air cooling	

provided by lithium-ion over lead-acid are the higher energy density based on a typical average couple voltage of 3.6 V relative to 2.0 V for the latter. This has an effect on both the specific energy and the energy density; therefore providing more energy for the available envelope allotted for the power source and decreasing the battery weight.

The current example of lithium-ion batteries replacing their lead-acid counterparts are two Li-ion multi-kWh product lines made by Ultralife Corporation. In this case, the economy of scale that is driving the need for replacement is based on the high number of land applications in which the upgraded power storage devices can be utilized. Among the land applications listed for Ultralife's units (e.g., URB0001, URB0003) are: in-line power backup, supplemental or reserve power, telecom and data centers, forklift trucks, personal transport, and medical devices [8]. These Li-ion batteries were designed as direct replacements for 1–5 kWh lead-acid batteries in 24 or 48 V applications. The battery design in this case favors maximum possible energy density for high-power devices with rapid charge capability. This means that first and foremost, power rating (W•kg^{-1}) and fast recharge (C/5–C/7; with 1C recharge possible but somewhat detrimental to battery's life) are considered. After the primary cell design factors, such as the cathode's chemistry, have been addressed, the system is then optimized for the optimum energy density. In the case of URB0001 and URB0003 batteries from Ultralife, the effective energy density at battery level is 73 and 103 Wh kg^{-1}, respectively [9]. These energy density values are modest when compared to the energy-optimized batteries (e.g., Patco Electronics' Li-ion batteries with lithium cobalt oxide cathodes that deliver >160 Wh kg^{-1}) [10] but for a power-optimized product, these values are competitive. The URB0001 and URB0003 batteries are based on lithium iron phosphate cathode, which enables fast charge/discharge rates while retaining good cycle life; see Chap. 1 of this book for more information on phosphate cathodes.

Apart from power and energy density, there are several other reasons why lead-acid battery replacement with lithium-ion batteries is worth considering. In addition to usually higher upper temperature limit (~20–30 °C difference, depending on the exact system details), coulombic charge versus discharge efficiency for lithium-ion is generally >97 % compared to 75 % for lead-acid. However, the cost of lithium-ion systems remains the most prohibitive factor in their competition with other battery chemistries, especially the lead-acid option. The high cost of raw materials on the cell level (3–4 times higher for Li-ion) is combined with the need for the expensive protective circuitry and battery management electronics which bring the total system cost to four times that of lead-acid battery of equivalent size. The comparatively high cost of Li-ion batteries per kWh versus lead-acid is at least partially offset by the longer replacement time frame (about 3–5 years difference), the lack of need for frequent maintenance and longer cycle life (>3,000 charge/discharge cycles for Li-ion batteries vs. ~300 cycles for lead-acid devices) [9, 11]. Given the factors described above, it is clear why the final battery chemistry selection requires thoughtful analysis of the predicted user profile.

In some cases of modern high-end land applications, lithium-ion batteries are much more than a replacement of an older technology. Such is the case of a Land Warrior US Army program, where many potential power sources and their hybrid systems have been considered. The Land Warrior program was launched in year 1994 and since then had undergone numerous scope, funding, and name changes [12–14]. In this text, the name "Land Warrior" is maintained due to the perseverance of this designation in battery products' technical literature. The related terms are "Nett Warrior" and the "Future Force Warrior." The Land Warrior system is meant to be deployed by infantry, and combat support soldiers, including rangers, airborne, air assault, as well as light and mechanized infantry soldiers.

Lithium-ion battery packs have emerged as some of the most successful products in terms of their field-readiness and consistency with the program goals. Since the inception, the rationale behind the program has been that on a modern battlefield, an integrated fighting system gives the infantry soldier enhanced tactical awareness, lethality, and survivability. To this end, the system package includes: the weapon system, helmet, computer, digital and voice

Table 4.5 Li-ion land warrior battery examples: Patco electronics

Part number	PB-LW-06	PB-LWH-01
Capacity	2.4 Ah	11.6 Ah
Energy	35 Wh	171 Wh
Energy density	118 Wh kg^{-1}	171 Wh kg^{-1}
Weight	0.295 kg (0.650 lb)	0.998 kg (2.2 lb)
Max. discharge current	2.0 A continuous	5.0 A continuous
Max. pulse discharge current	4.0 A for 5 s	7.5 A for 5 s
Charging	1.8 A to 16.8 V in a temperature range of 0 to 45 °C	3.0 A to 16.8 V in a temperature range of 0 to 45 °C
Voltage range	10.0–16.8 V	
Nominal voltage	14.8 V	
Cycle life	>300 cycles at C/5 and 100 % depth of discharge; 80 % of retained initial capacity	
Operating temperature	−20 to 60 °C	−30 to 60 °C
Storage temperature	−20 °C to 60 °C	
Self-discharge	<3 % per month	
Exterior/housing	Hard plastic	

communications, positional and navigation system, protective clothing, and other individual equipment. Considering the number of power and energy demanding electronic devices that are part of the package and their combined weight, unsurprisingly, the weight and volume savings are sought for all components including the power sources.

The specifications shown in Table 4.5 represent two different lithium-ion Land Warrior batteries made by Patco Electronics, Inc. The units shown in Table 4.5 are from the two opposing ends of the range regarding their capacity. These batteries are used in military training systems, backup server power, and ruggedized portable electronics [12–14].

The variety of lithium-ion batteries with different specifications sold to meet the needs of the Land Warrior application suggests that the universal power source has not been developed yet. There is a mismatch between the fast growth rate of performance of electronic devices accompanied by their power demands and the incremental improvements made to Li-ion batteries.

The Army Research Laboratory/Army Communications Electronics Command (ARL/CECOM) in 2002 defined three separate power regimes for the Land Warrior application: 20 W average with 50 W peak, 100 W average with 200 W peak, and 1–5 kW high power drain applications. The batteries illustrated in the Table 4.5 fall within the first range (PB-LW-06) and partially fulfill the requirements of the second regime (PB-LWH-01). The relatively short guaranteed cycle life of these batteries (300 cycles) is based on the expectation of frequent deep discharges (100 %), which are technically possible but very detrimental to the cathode material (in this case, $LiCoO_2$; see Chap. 1 for more information on these cathodes).

Another important facet of this high-end military application is safety, since the wearable energy source is integrated closely within the soldier's wardrobe and it should not pose a threat, e.g., when struck with a projectile. The high-end, soldier-wearable batteries are subject to very stringent packaging requirements, where the battery casing should be capable of resisting violent reaction upon being breached, such as by being punctured or crushed. Ruggedized construction and the rigorous safety demands are the common characteristics of all the military applications.

4.1.2 Human Implantable Lithium Batteries

Human implantable batteries belong to a unique class of land applications that defines yet another set of strict requirements. There is the necessity for reliability, with no tolerance for the risk of failure because in the case of fully implantable devices, a surgery is required each time the battery needs to be replaced. The cells must be small and provide reliability to deliver adequate cycle life to gain the most out of the procedure to implant them. Depending on the type of implant, the surgery is sometimes followed by a lengthy recovery period. Additionally, due to the direct contact with human tissues, the cell technology for this application must be well-sealed, especially with the use of a nonaqueous electrolyte.

An example of the human implantable Li-ion cell is the model used for powering of the fully-implantable Carina® middle ear hearing aid by Otologics, LLC [15–18]. The Otologics implant contains the electronic capsule, subcutaneous microphone, and transducer. Inside the electronic capsule, the battery, magnet, digital processor, and the radiofrequency coil are contained. Signals picked up from the microphone are relayed to the transducer that has a tip mounted in a laser-drilled hole in the body of the anvil (one of the ossicles in the middle ear). As the transducer vibrates, a probe transfers these vibrations onto the anvil. The transformation of sound waves to vibrations is aided by the programming system, which communicates with the rest of the implant via wireless connection. All the operations and the communication processes of the implant are powered by a rechargeable Li-ion battery.

The battery charger and the remote control are the external parts performing the supporting functions to the implant. The remote control placed by the patient outside of the skull and directly over the implanted coil turns the implant on and off and adjusts the volume. The charger system includes a base station, charging coil, and charger body. The charging of the fully implanted Li-ion battery (consisting of a single Li-ion cell with basic electronics) is done via inductive coupling, where the receiving coil is directly connected to the battery and the sending coil is part of the external battery charger. The battery is designed to work during the

Table 4.6 Carina® lithium-ion battery specifications

Battery characteristic	Value
Capacity	0.05 Ah (Nameplate)
Specific energy	60 Wh kg^{-1}
Energy density	175 Wh L^{-1}
Weight	0.004 kg
Max. discharge current	0.25 A
Max. pulse discharge current	0.5 A
Cycle life	>2,000 cycles
Operating temperature range	−20 to 60 °C
Other characteristics	Biocompatible (sterilized, stainless steel case)

day between nightly charging sessions that require about 1 h. The hearing aid still receives and transmits signals during charging. Any preexisting residual hearing is expected to be retained even at the end of the battery's life.

The initial generations of lithium-ion batteries for early iterations of Carina® were made by Yardney Technical Products, Inc. with the eventual target of technology transfer to Otologics. The original lithium-ion cell model is still available from Yardney under the product name of NCP0.05–1. Properties of this battery are listed in Table 4.6. The Carina® hearing systems have received the CE Mark for clinical use in Europe, and advanced trials on multiple patients (more than 50) were well underway when Otologics filed for bankruptcy in July of 2012, with a potential asset buyer being actively sought after. Currently, there are two implantable hearing aids comparable to Carina® that are in similar stage of advancement: Vibrant Soundbridge® (MED-EL, Austria)—a partially-implantable device, and Esteem® (Envoy Medical Corporation, USA), which is a fully implantable unit.

From the technological standpoint, the fully implantable hearing aids represent an undeniable advancement. However, there are several factors that constitute barriers against their wide commercialization. First, the placement of these implants requires a general anesthetic and approximately a 2-h operation. Second, the price has been reported at $12,000.00 and $15,000.00 per unit (2010), which does not include the surgery and related medical costs. Then, there is a question of the insurance coverage, which is currently problematic for the fully implantable hearing aids, especially since the partially implantable or external hearing aids are less invasive and in most cases perform their function to a similar extent. However, the advances in the medical prostheses of various types is expected to provide the increasing opportunity for development and use of advanced implantable power sources, including the high-end lithium-ion batteries (Table 4.7) [19]. Two likely potential medical applications for lithium-ion technology in the near future (5–10 years from today) are: "bionic eyes" [20] and colonoscopy pill probes/cameras [21].

Table 4.7 Lithium-ion battery-powered medical implants

Battery manufacturer	Product name	Implant function
Medtronic, Inc. [22]	RestoreUltra® (Implant brand name)	Spinal cord stimulation and muscular pain management
Greatbatch, Inc. [23]	Xcellion™ (Battery product line)	Implantable neuromodulation, orthopedic and cardiac rhythm management devices
Quallion, LLC [24]	QL0003I, QL0020B, QL0200I-A (Batteries)	Medical implants, sensors: cochlear implants, spinal cord stimulators, glucose sensors, and cardiac rhythm management devices
EaglePicher medical power, LLC [25]	CONTEGO, Micro battery (Battery product lines)	Neurostimulation, middle ear, and cochlear implants, drug pumps and body fluid pumps, telemetry (monitoring blood pressure, temperature, etc.)

4.1.3 Aerial

Aerial applications requiring advanced lithium-ion batteries consist of top-level military aircraft, both manned and unmanned. With aerial platforms, advantages of Li-ion technology include volume and weight savings as well as the extended operating and storage temperature windows. Safety considerations are based on the fact that the battery should contribute the smallest additional risk to the already complex and expensive vehicle. In many aerial applications, lead-acid, nickel-cadmium (Ni-Cd), and nickel-metal hydride (Ni-MH) batteries were used in the past but have been replaced with lithium-ion devices due to their technological advantages.

Lithium-ion technology when appropriately optimized, can deliver up to four times the energy of lead-acid (Table 4.8). The energy gains are smaller in case of the Ni-Cd or Ni-MH upgrades to Li-ion, but there are advancements in power capability, accompanied by the temperature range widening and lower maintenance requirements. For an aerial platform, weight and volume gains are of serious consideration since this affects the amount of fuel used for a mission: the lithium-ion battery is an asset, satisfying the performance requirements, while being smaller and lighter than the obsolete units based on older technologies.

Batteries in some aerial applications are required to survive operation and storage at temperatures of −40 °C and above 70 °C without significant energy loss. Such conditions would typically induce a significant loss of energy or total failure in "standard" commercial Li-ion batteries. A common challenge for high-end aircraft batteries is undergoing the rapid temperature change from ambient conditions at ground level to elevated temperatures of the aerial trajectory to reach maximum altitudes. The entire process is accomplished in seconds and capable of causing significant temperature shock to the battery. Temperature shock is poorly tolerated

Table 4.8 Li-ion specifications versus Ni-MH, Ni-Cd, and lead-acid batteries

Cell ID	Capacity	Continuous discharge	Pulse disch.	Operating temp.[a]	Energy density	Spec. energy
	Ah	C-rate	°C	Wh L^{-1}	Wh kg^{-1}	
NCP55-1 (Yardney, aerial application)	50	4C	8C	−40 to +70	265	112.5
Ni-Cd (Literature)[b]	N/A	5C	–	−20 to +50	110	56
HHR210AA/B panasonic (Ni-MH) [26]	2	2C	–	0 to +45	287.8	82.8
PS-260, lead-acid power-sonic corp. [27]	6	1.3C	4.3C	−20 to +50	70.4	29.7

[a]For NCP55-1, the operating temperature range represents both charge and discharge; for Panasonic and Power-Sonic batteries, the narrower (charge) range was given for the realistic comparison

[b]Unlike the NCP55-1 (Yardney) cell, the Panasonic and Power-Sonic cells were not made specifically for the high-end aerial applications; the latter represent the advanced, commercial state of the art; the table comparison should be therefore treated as an approximation

by commercial batteries due to their inability to resist the thermal expansions or contractions of their component materials. These mechanical battery failures that result from thermal shock are usually localized at the cell enclosure, especially near electrode tab feed-throughs. In high-end aerial batteries, the mechanical hazards are addressed by the cell design enhancements such as more robust feed-throughs and mating several electrodes at the carefully selected contact points.

Another type of thermal shock hazard is related to the possibility of lithium-plating on the anode plates' surfaces [28]. This chemical battery failure is the consequence of highly impaired ionic transport occurring in the rapidly cooled electrolyte solution. Combinations of specialized electrolyte additives have been developed to preserve the electrolyte's performance at low temperatures as well as improve the stability of electrolyte solution at elevated temperatures [29]. The low temperature performance challenges are tied to decreased ionic conductivity and increased viscosity experienced by "standard" electrolytes at low temperatures and the low temperature additives are developed to counteract these phenomena. The high temperature electrolyte enhancements are based on the principle of electrodes' passivation layers and strengthened interface between electrolyte, electrodes' surfaces, and separator positively contributing to electrolyte's stability throughout the extended temperature window. The presence of multiple additives and the robust electrodes' passivation layers are expected to negligibly affect the internal resistance of the cell, so that the full discharge energy is still available for use. For the discussion of the relationship between the internal cell's resistance and the discharge energy and rate, see Text Box 3.6 in the Chap. 3 of this book.

Another important characteristic that is necessary for aerial vehicle batteries is their ability to withstand stresses related to mechanical shock and vibrations through their projected lifetime. The batteries must not fail as they are an integral portion of the system providing load support to the engine for start-up power and assisting hydraulics in specific maneuvers [30].

Aerial vehicle examples that benefit from advanced lithium-ion technology include the B-2 Stealth and Global Hawk. These are high-end applications that cannot suffer from battery failure and require responses to rapid environment changes. The latest of battery applications in other U.S. fighter jets include several versions of FA-18 planes [31] and the most recent addition to the arsenal, the F-35 Fighter. Because of the significant g-forces experienced by these aircrafts and their wide operating temperature windows, the toll taken on the batteries accumulates quickly and easily, leaving no place for the design or fabrication shortcomings. The most advanced inclusions into the battery design pertain to the handling of thermal transfer both within the battery and between battery and the surrounding aircraft components. The facile thermal transfer prevents heat from being "bottle-necked" within the battery, which affects both performance and safety aspects of the advanced batteries' operation. More detailed discussion of thermal modeling and the related cell design modifications are found in Chap. 5 of this book.

4.1.4 Space

Soon after their successful market introduction in the early 1990s, lithium-ion batteries became an attractive choice for powering various space missions. As with other high-end applications, the weight and volume savings provided the most powerful incentives to use Li-ion batteries instead of the alternative battery chemistries [32, 33]. The gravimetric and volumetric characteristics are of primary importance in a craft that must launch, break free of Earth's atmosphere and carry the payload into space. Low life cycle cost and minimal maintenance needs were other factors contributing to the selection of Li-ion technology-based power system. In these applications, the low cost is achieved by extremely long cycle life [34] demonstrated by lithium-ion units (>30,000 cycles at partial depth of discharge for selected applications). The diverse group of space applications is comprised of: planetary landers, planetary rovers, planetary orbiters, earth orbiting spacecraft (geosynchronous earth orbit—GEO and low earth orbit—LEO), astronaut equipment and, recently, crew exploration vehicles [32, 35]. The major end user in the case of space applications is the National Aeronautics and Space Administration (NASA) supported by various Original Equipment Manufacturers (OEMs) such as Lockheed Martin Corporation and Boeing Corporation [36]. In addition to NASA, other major users include the US Air Force and the communications satellite industry.

There are several examples of NASA space missions powered by lithium-ion batteries. Among these were the two prominent Mars Exploration Rover (MER) programs, pertaining to rovers launched in the summer of 2003 to survey the surface of Mars, nicknamed Spirit and Opportunity [37, 38]. Both rovers had lithium-ion batteries working in unison with a solar panel charging system. Each rover was equipped with a battery assembly unit, consisting of two 8-cell, 10 Ah (nameplate capacity of 8 Ah) strings connected in parallel, manufactured by Yardney Technical Products, Inc. The battery was designed to typically deliver 40–50 % depth of discharge (DOD) each sol (martian day).

The MER missions imposed several operational requirements on the batteries, such as:

(1) Maintaining the operating voltage of 24–36 V.
(2) Providing high energy (e.g., 220 Wh) to support the launch activities.
(3) Correction of any fault-induced attitude deviation during the cruise period (e.g., 160 Wh).
(4) Providing sufficient energy for surface operations ($\geq$283 Wh sol^{-1} at 0 °C).
(5) Delivering sufficiently long cycle life ($\geq$270 cycles at 50 % DOD and/or 90 sols of operation).
(6) Maintaining the ability to support multiple pulses of 30 A for 50 ms, including low temperature (e.g., −20 °C) pulses.
(7) The battery should successfully charge and discharge between −20 and 30 °C throughout the length of the entire mission [39].

Spirit and Opportunity were both sent to Mars for 90-sol missions that were completed successfully. The post-mission traverse of Spirit was terminated when the wheels of the vehicle became trapped in the soft soil [40, 41]. The total distance traveled by Spirit was over 7.7 km and the entire mission of this MER rover lasted 2,555 sols. As of June 27, 2013, the Opportunity rover has been continuing the amazing >37-km journey at 3349 sols and counting.

In terms of their performance evaluation, lithium-ion batteries used in Spirit and Opportunity rovers have far exceeded their mission objectives. Some power system challenges that became apparent during the operation of the twin MER rovers were related to the martian dust coating the solar panels, so that they became inefficient in recharging the batteries. Occasionally, dust would be blown away and the recharging process could resume but this environmental factor introduced another variable on the already complex mission. The future power systems for Mars surface vehicles would require modifications to the principal power source in order to gain more independence from the environmental conditions.

As a follow-up to the success of the original Mars Rovers, NASA sponsored and developed the Mars Science Laboratory (MSL) named Curiosity. This vehicle had the high-end lithium-ion battery providing backup power to the U.S. Department of Energy's radioisotope energy generator. The MSL was launched on November 26, 2011 and landed on the surface of Mars on August 5, 2012. The anticipated duration of Curiosity's prime mission is one Mars year (669 Sols).

The continued use of Li-ion batteries in space applications and the increased mission duration projections are based on the past performance of the delivered batteries combined with the array of rigorous space application qualification tests. Typically, the rover battery candidates undergo performance and abuse tests on the battery and cell levels to establish an engineering evaluation database, spanning

Table 4.9 Li-ion space battery qualification test: random vibration

Input: Vibration frequency, Hz	Output: hardware vibration response (qualification level), $G^2\ Hz^{-1}$
20–80 80–450	+6 Db per octave 0.08
450–2,000	– 6 Db per octave
Overall	7.8 G^{a}_{rms}

[a]G_{rms}—root-mean-square acceleration

across data originating from multiple cell/battery candidates. The performance tests include physical characterization (dimensions and weight), electrochemical characterization (open circuit voltage, closed circuit voltage, and capacity checks), rate capability (capacities at different charge/discharge rates and different temperatures), and vacuum leak checks. The examples of abuse battery tests include overcharge, internal and external short, crush, drop, pyroshock, and vibration. In the course of testing, the cell and battery level protection switches are also identified and evaluated. The "generic" (done across multiple applications) abuse tests, such as overcharge, short, and crush-testing are described in more detail in Chap. 5 of this book. The shock and vibration tests and allowable levels are more specific to the actual space (and aerospace) applications and are discussed in more detail below.

The rover batteries (e.g., Spirit, Opportunity, and Curiosity) all underwent the vibration tests applied to the battery in x, y, and z directions for 2 min during the qualification tests (the vibration time varies between application and the level of testing). Random vibrations (e.g., derived from launch vehicle vibration profiles during past missions) were applied at the battery-mounting frame interface and the battery hardware response (also in the form of vibrations) was measured [42, 43]. The batteries were powered-on during vibration testing, simulating the requirement of operation during launch. The typical random vibration testing profile for rover battery qualification is given in Table 4.9. The purpose of the random vibration testing is to estimate the potential battery behavior (or damage, if the battery grossly fails the test) during the launch conditions, which next to landing, can be the most detrimental to the state of health of the battery.

Pyroshock is the decaying, oscillatory response of a structure to high-amplitude and high-frequency mechanical excitation. The frequencies that comprise this oscillatory response can extend to thousands of Hertz and beyond [44]. For example, the space vehicle launch or jettisoning of the used rocket stages and other similar events are known to initiate the pyroshock. In the past, large components of the space vehicles typically had low resonant frequencies, which effectively isolated them from pyroshock. However, over the years, electrical components (such as battery electronics) have become increasingly more miniature. Because of this miniaturization, the mechanical resonant frequencies of these components have increased; making them susceptible to damage by pyroshock and thus, the component testing for pyroshock response became necessary (Table 4.10).

Table 4.10 Li-ion space battery qualification test: pyroshock exposure

Input: vibration frequency, Hz	Output: hardware shock response (qualification level)
100	20 G
100–1,600	+10 Db per octave
1,600–10,000	2,000 G

Table 4.11 Li-ion rover battery qualification: thermal vacuum[a] requirements

Mission segment	Qualification temperatures (°C)
During cruise	−30–40
Charge on mars	−10–40
Discharge on mars	−30–40

[a]The pressure for the test is reduced from atmospheric to 10^{-5} Torr or less

The above requirements constitute a small portion of the expansive suite of testing methods that ensure the mission endurance without compromising the battery's/cell's integrity. For Li-ion cells maintaining integrity means not allowing any electrolyte leakages potentially induced under harsh conditions, such as acceleration and/or vibrations associated with the launch, cruise, and landing of the vehicle. Cell integrity is typically ensured by using laser-welded stainless steel cell cases and utilizing glass-to-metal seals around the cell terminals. The successful cell sealing is of the utmost importance since the space environment, i.e., vacuum, can rapidly evaporate the electrolyte solvents and degrade the cell performance to null. For this reason, rigorous leak and vacuum integrity testing must be done on Li-ion cells and batteries at various temperature conditions, as illustrated in Table 4.11.

Aside from the integrity of the cell case, shock and/or vibration conditions could lead to the misalignment of the electrode pairs and/or separator sheets or disrupt the electrode stacks' connections to the terminal feed-throughs. Therefore, the electrodes' stacks must be engineered in such a way that there is adequate internal stack pressure to prevent slippage under the most extreme vibration and shock conditions. This aspect of the cell design requires an understanding of the volume changes occurring within the electrodes' stack during the repeated charge and discharge cycles, especially prominent upon formation cycles and intimately related to the electrodes' chemistry.

Another chemistry-related challenge of space travel relates to maintaining the electrolyte solution integrity during voyage. Custom solvent mixtures and lithium salts are used in space-rated batteries, with melting points lower than the temperatures anticipated in the battery compartment during flight. Solidification of the liquid electrolyte would produce inhomogeneity in the cells and therefore preclude their efficient performance at the time of landing.

The above considerations illustrate the unique set of challenges that space travel imposes on all power sources. So far, lithium-ion batteries have been very successful in meeting and exceeding the space missions' requirements but with each new mission, the expectations are increased. In addition to the desired continuous improvements to Li-ion technology, the cooperating power sources, such as the radioisotope energy generators, are expected to evolve.

4.1.5 Sea

Marine high-end battery applications include above water systems, such as electric ships, as well as submersibles, both manned and unmanned. These categories break down into a wide variety of subcategories based on mission needs, size, location for their utilization, etc. These applications predominantly consist of military platforms; however, commercial needs also exist, such as providing high-energy storage systems in underwater exploratory vehicles.

The above water batteries are typically required to operate within a wider window of ambient temperatures than their undersea counterparts and typically have fewer constraints on their volume and weight. In undersea vehicles, one of the greatest challenges is maintaining appropriate buoyancy; however, water provides a moderating influence on the battery's operating temperature window. The exact choice of the thermal management method (forced air, circulating coolant, or convection) is dictated by the type of vessel and the vessel's site of operation.

In selected cases described in this section, batteries provide the main source of propulsion of the vehicle. However, hybrid systems are becoming more popular due to their specifications' flexibility and variety of different power source devices available for hybridization [45]. In hybrid power systems, batteries can be coupled with fuel cells, capacitors, supercapacitors, hydrocarbon fuel engines, or diesel generators. The role of the battery in such hybrids is typically to assist in load-balancing and/or to provide more efficiency in meeting power requirements [46]. More detailed discussion of hybrid power systems is included in Chap. 6 of this book.

Electric ships constitute a diverse and rapidly growing battery application, where fully electric operation might be required at times to provide stealth propulsion mode (military applications) or reduce CO_2 and NO_x gaseous emissions (primarily sought after in civilian applications) [47, 48]. Alternatively, a battery could be working in conjunction with the boat's diesel engine, storing power produced by the generator and providing auxiliary propulsion power when required.

A recent commercial example of the large ship having a hybrid-electric power supply system is the car carrier, the Emerald Ace, jointly developed by Panasonic Corporation, Mitsui O.S.K. Lines and Mitsubishi Heavy Industries, unveiled in June 2012 [49]. The carrier was designed to achieve zero emissions while in harbor. The hybrid power supply system consisted of the 160 kW solar generation module, the 2.2 MWh lithium-ion battery module, and the diesel power generator. In addition to this specific example, the entire class of hybrid-powered cargo ships—the liquefied natural gas (LNG) carriers are being developed and implemented worldwide [50]. Several factors are driving the electrification of civilian ships. The commercial ships in North American waters are required to burn expensive low-sulfur oil. Another incentive is provided by the restrictions on greenhouse gas emissions by commercial ships phased in by United Nations' International Maritime Organization. The above factors lead to the increasing electrification of commercial ships.

In the noncivilian marine applications, the U.S. Navy has been using batteries for decades on multiple vessel classes to support various functions: assist in propulsion, power on-board electronics, sensors, communication devices, and weapons. As evidenced in the Naval Power Systems Technology Development Roadmap, released in 2013, the steadily increasing power supply requirements, economical factors (cost of fuel) and the increasing mission duration times provide a powerful push for the increased use of electrical power on board of Navy's vessels. Due to their high energy density, high rate capability, and long cycle life, Lithium-ion batteries are one of the leading candidates for powering the future electric ships, but first their past safety issues need to be fully addressed. Chapter 5 of this book contains the discussion of Li-ion technology safety challenges and the ways to address them.

For large naval platforms of all types and application sites, hybrid systems are being developed, but the system size must be large enough to achieve the practical energy gain versus available weight and volume ratio. The most current examples of the US Navy ships powered by the hybrid propulsion systems are the USS America (LHA-6) and the USS Tripoli (LHA-7), which are part of the America-class amphibious assault ships group. Their predecessor was USS Makin Island (LHD-8), a Wasp-class amphibious assault ship. The hybrid-electric propulsion systems used on these hybrid-electric ships consist of a gas turbine engine as well as an electric motor and diesel generator. The electric motors can help propel the ship at speeds up to around 12 knots and the generator can help produce electricity for the ship. When it comes to traveling at speeds greater than 12 knots or so, the ship then relies upon the gas turbine engine. At the same time, the generators can provide on-board power for many of the ships systems, such as sensors, weapons, and other electronics.

The Navy submersible vehicles that have benefited from the use of lithium-ion batteries encompass both manned and unmanned crafts. An example of a large naval manned vehicle that incorporated lithium-ion batteries was the Advanced Seal Delivery System (ASDS), currently on hiatus. One of the main challenges facing submarine batteries is the heat dissipation. Large battery banks (megawatt scale) in submersibles are often packed in pressure vessels; therefore, the means to cool the battery are often severely limited to convection across the battery rack and through the shell. Submersible vehicles that are propelled using exclusively battery power (e.g., ASDS) need to expend the battery's energy on propulsion and on supporting the mission requirements.

Typically, the propulsion requirements determine the operating voltage (e.g., the ASDS average battery voltage ~250 V). Due to the high voltage needs, lithium-ion technology is very attractive for these applications, since fewer cells (3.6 V per Li-ion cell) are required in series to meet the requirement. Additionally, the energy density furnished by Li-ion batteries exceeds that of others on a small scale application [51].

4.2 Quality and Reliability for High-End Lithium-ion Technology

4.2.1 Quality

Quality assurance in high-end lithium-ion applications is crucial. As described in the previous parts of this chapter, there is no room for failure and in many cases, the possibility of repair cannot be considered. Therefore, at the outset of choosing the battery technology for a specific application, a comprehensive quality system is necessary. The battery design considerations from the quality standpoint include, among others, the battery chemistry (inclusive of the anode, cathode, and their loading, the separator, electrolyte), the cell geometry and feed-throughs, cell pack size and design.

The two most prevalent quality management systems used by manufacturers of diverse products, including Li-ion (high-end and commercial) batteries are developed by the International Organization for Standardization (ISO) and the American National Standards Institute (ANSI). Aside from stating the most appropriate testing and/or processing procedures, these standards are auditable, well documented to readily verify adherence.

The compliance with any extensive quality system requires a long "paper trail" but is necessary to assure the pedigree of the battery components, starting from raw electrodes' materials and ending with the battery casing and electronics. In case of the high-end batteries, the complete traceability and pedigree of their components are among the standard end user requirements. There is an inherent value in the existence of this vast database, which proved useful in the past occurrences of nonconformance, battery failure or simply when the battery design was being reevaluated and upgraded.

In addition to the industry-wide ISO and ANSI standards, other, application-specific quality certification systems are often used. For the Aerospace-certified products, the AS9100 standards are often preferred and some large Original Equipment Manufacturers (OEMs) have developed their own quality systems (e.g., Boeing North America—Boeing D6-82479).

Typically, as the result of quality control system implementation, procedures are in place to establish and define the methods of controlling the progress of changes in hardware and processes. In addition, these procedures designate responsibility for directing, documenting, processing, and recording the changes. This assures correct detail preparation and revision control of all applicable engineering drawings, specifications, supplemental and quality assurance provisions.

4.2.2 Reliability

It has been over two decades since lithium-ion batteries were introduced to the marketplace (circa 1991). After about 8 years of proven and predictable performance, this technology began to venture into the high reliability markets. The successful entrance into these markets was achieved by combining proven chemistry, premium materials, state of the art equipment and processes, an exhaustive quality system. This entrance into high reliability applications arose, primarily, from the increasing energy demands of today's electronics and the demonstrated reliability of the technology.

Reliability analysis is an engineering tool commonly used in high-end products' design and manufacturing to provide the statistical estimate of the system's capability to function without failure while exceeding the requirements placed upon the product. For advanced battery systems, reliability analysis is an integral part of the battery development process along with the failure modes and effects analyses. The hazard analysis assures safely meeting mission objectives within the required battery lifetime. A thorough reliability analysis takes into account every component in the system using their individual lifetime estimates for operation and storage in particular conditions, such as temperature, pressure, exposure to elements, such as moisture and radiation, and other possible conditions that may have an influence on performance and lifetime. Through testing of parts, such as electrical components, e.g., DC/DC converters, processors, etc., a typical lifetime is determined with a high confidence level.

The reliability analysis' output is often expressed as a percentage, which, given the specifically defined conditions, provides the confidence level that the system will not fail. Another commonly used output of the reliability analysis is the Mean Time Between Failure (MTBF), which indicates how long a system is expected to function without the need for repair [52].

Based on their need for reliability, the customers may request in the product specification that the system has a strictly defined high confidence level (e.g., 98 %). The predetermined level of confidence impacts the choice of product's components and selection between different available component grades, balancing cost with quality. The component grade selection is also dependent on how the system is designed, i.e., how many units of a specific component are used and if there is any component redundancy. Equations that are used to calculate the reliability factors are as follows:

$$R(t) = e^{-\lambda t} \tag{4.1}$$

where:

- $R(t)$ is the reliability factor as a function of the minimum operation duration without failure;
- λ is the failure rate, i.e., number of failures experienced in certain time of operation of the system
- t is the mission duration, expressed in units of time (e.g., hours)

$$\mathrm{MTBF} = \frac{1}{\lambda} \tag{4.2}$$

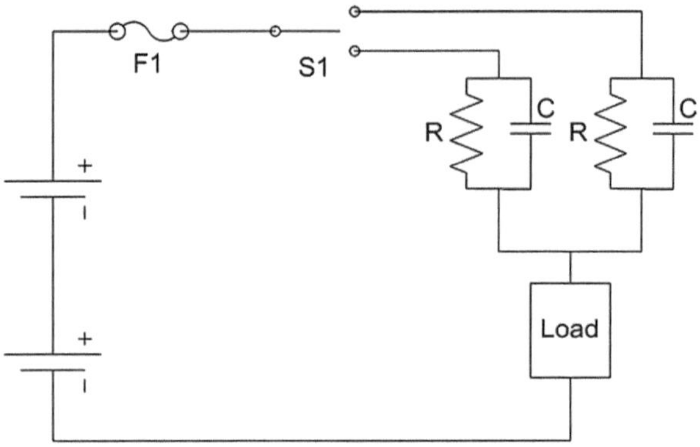

Fig. 4.5 Simple circuit showing two identical and independent (redundant) capacitor/resistor paths in an electronic circuit with a load, two cell battery (in series), a fuse and a single pole double throw (SPDT) switch.

Text Box 4.1 An example of MTBF calculation

If a product specification requires the system reliability of 99.0 % for mission duration of 200 h, then the MTBF can be calculated as follows:

$$R(t) = 0.990 = e^{-\left(\frac{1}{\mathrm{MTBF}}\right)200}, \text{ therefore } \ln(0.990) = -\frac{200}{\mathrm{MTBF}}$$

The calculated MTBF $\approx$ 19,900 h for the entire system

Typically, redundancy is an entirely independent portion of the system that can be switched on in the event of a failure of the identical, originally used counterpart. The use of redundancy increases the reliability of the system. An example of a simple circuit that has a redundant resistor/capacitor portion is shown in Fig. 4.5.

In their concerted efforts to ensure reliability, the high-end battery manufacturers are also exceedingly selective in their choice of raw materials. While the price difference between the selected high-end raw materials can be five to six times more than their similar "low-cost" counterparts, the extensive performance database and unparalleled performance trump the cost factor in the case of high-end applications. In addition to selecting premium materials, multiple viability/quality checks are incorporated into the cell and battery production process. Finally, in reviewing the reliability of lithium-ion cells in relation to a prospective high-end application, extensive laboratory testing is done that covers a range of temperatures, rates, depths of discharge, etc. Typically, the manufacturers conduct several (from 1 or 2 to 15 cycles) pre-ship cycles. Factors such as columbic efficiency, energy efficiency, and self-discharge rate are calculated for each high-end cell.

4.3 Conclusion

The common denominator for the group of high-end battery applications is the prevalence of performance metrics over cost per unit. In these applications, the deliverable hardware is often low-volume, sometimes unique and extremely specialized to match the end user's specific needs. For this reason, the high-end applications have always been "paving the way" for the commercial applications, implementing the game-changing concepts and reducing these into practice well before the commercial battery industry. However, making the transition into the commercial battery production is by no means trivial, as addressing the cost factors and the technical requirements of mass production processes necessitate different approach and the relevant expertise.

References

1. http://www.everything-robotic.com/2011/10/emerging-underwater-businesses-three.html. Accessed on June 25, 2013
2. Reddy TB, Linden D, eds (2010) Linden's handbook of batteries 4th edition. McGraw-Hill, New York
3. http://yardney.com/Lithion/lithion.html. Accessed on June 25, 2013
4. http://www.hitachi-ve.co.jp/en/products/spec/index.html. Accessed on June 25, 2013
5. www.gsyuasa-lp.com/download/file/fid/152. Accessed on June 25, 2013
6. http://www.panasonic.com/industrial/batteries-oem/oem/lithium-ion.aspx. Accessed on June 25, 2013
7. http://www.samsungsdi.com/battery/prismatic-ICP103450.jsp. Accessed on June 25, 2013
8. http://ultralifecorporation.com/be-military/products/military-rechargeable/. Accessed on June 25, 2013
9. http://ultralifecorporation.com/downloads/. Accessed on June 25, 2013
10. http://www.patcoelectronics.com/products/rechargeable-lithium-ion-batteries/. Accessed on June 25, 2013
11. http://www.power-sonic.com/sealed_batteries.php. Accessed on June 25, 2013
12. Committee on Soldier Power/Energy Systems Board on Army Science and Technology Division on Engineering and Physical Sciences, National Research Council of the National Academies (2004) Meeting the Energy Needs of Future Warriors. The National Academic Press, Washington DC
13. http://www.popularmechanics.com/technology/military/4215715. Accessed on June 25, 2013
14. http://www.army-technology.com/projects/land_warrior/. Accessed on June 25, 2013
15. http://hearinglosshelp.com/weblog/526.php. Accessed on June 25, 2013
16. Bruschini L, Forli F, Santoro A, Bruschini P, Berrettini, S (2009) Fully implantable Otologics MET Carina™ device for the treatment of sensorineural hearing loss. Preliminary surgical and clinical results. Acta Otorhinolaryngol Ital 29:79–85
17. Luers, JC, Beutner D, Huettenbrink, KB (2011) Implantable hearing aids. HNO 59:980-987. doi:10.1007/s00106-011-2402-0
18. Flint PW et al. (2010) Cummings Otolaryngology - Head and Neck Surgery, 5th edition. Mosby – Elsevier, Philadelphia, PA
19. http://www.nerac.com/nerac_insights.php?category=articles&id=151. Accessed on June 25, 2013

20. http://engineering.illinois.edu/news/2013/02/28/stretchable-battery-flexible-circuits. Accessed on June 25, 2013
21. Valdastri, P, Simi, M, Webster, RJ (2012) Advanced Technologies for Gastrointestinal Endoscopy. Annu Rev Biomed Eng 14:397–429. doi: 10.1146/annurev-bioeng-071811-150006
22. Medtronic (2007) RestoreULTRA™ Implant Manual. No. 37712
23. Xcellion™ Rechargeable Batteries Specifications Sheet. Accessed on June 25, 2013. http://www.greatbatchmedical.com/assets/products/Xcellion%20Rechargeable%20Batteries.pdf
24. http://www.quallion.com/sub-sp-main.asp. Accessed on June 25, 2013
25. EaglePicher Medical Power. Products and Services Catalog 2010. http://www.eaglepicher.com/images/Medical/EaglePicher%20Medical%20Brochure%202010.pdf. Accessed on June 25, 2013
26. http://www.panasonic.com/industrial/batteries-oem/oem/nickel-metal-hydride.aspx. Accessed on June 25, 2013
27. http://www.power-sonic.com/images/powersonic/sla_batteries/ps_psg_series/2volt/PS-260_11_Feb_21.pdf. Accessed on June 25, 2013
28. Smart MC, Ratnakumar BV (2011) Effects of Electrolyte Composition on Lithium Plating in Lithium-ion Cells. J Electrochem Soc 158:A379–A389. doi:10.1149/1.3544439
29. Gulbinska M, Moore G, Santee S, Lucht B, Puglia F (2011) Comprehensive improvements in Li-ion batteries for demanding applications. J Power Sources 196:2899–2904. doi:10.1016/j.jpowsour.2010.10.087
30. MIL-STD-810G, Department of Defense Test Method Standard
31. Fleener W (1997) Aircraft Main Battery Development Program for the F/A-18E/F Super Hornet Aircraft. SAE Technical Paper 971220. doi:10.4271/971220
32. Surampudi S, Halpert G, Marsh RA, James R (1998) Lithium-Ion Batteries for Aerospace Applications. 9th International Li Battery Meeting, Edinburgh, Scotland
33. Marsh RA, Vukson S, Surampudi S, Ratnakumar BV, Smart MC, Manzo M, Dalton PJ (2001) Li ion batteries for aerospace applications. J Power Sources 97:25–27. doi:10.1016/S0378-7753(01)00584-5
34. Smart MC, Ratnakumar BV, Whitcanack LD, Puglia FJ, Santee S and Gitzendanner R (2010) Life Verification of Large Capacity Yardney Li-ion Cells and Batteries in Support of NASA Missions. Int J Energy Res 34:116–132. doi: 10.1002/er.1653
35. http://www.nasa.gov/pdf/396773main_mf18_ouellette.pdf. Accessed on June 25, 2013
36. http://osbp.nasa.gov/docs/MARS%20OSB_CS55_FINAL_LO=TAGGED.pdf. Accessed on June 25, 2013
37. http://www.jpl.nasa.gov/news/fact_sheets/mars03rovers.pdf. Accessed on June 25, 2013
38. http://www.universetoday.com/96553/quick-and-curious-facts-about-the-mars-science-laboratory-mission/. Accessed on June 25, 2013
39. Ratnakumar BV, Smart MC, Kindler A, Frank H, Ewell R, Surampudi S (2003) Lithium batteries for aerospace applications: 2003 Mars Exploration Rover. J Power Sources 119:906–910. doi:10.1016/S0378-7753(03)00220-9
40. http://marsrover.nasa.gov/mission/traverse_maps.html. Accessed on June 25, 2013
41. http://www.space.com/18766-spirit-rover.html. Accessed on June 25, 2013
42. McKissock B, Loyselle P, Vogel E (2008) Guidelines on Lithium-ion Battery Use in Space Applications. NASA Engineering and Safety Center Technical Report NASA/TM-2009-215751, NESC-RP-08-75/06-069-I
43. Manzo M, Brewer JC, Bugga BV, Darcy EC, Jeevarajan J (2010) NASA Aerospace Flight Battery Program. Generic Safety, Handling and Qualification Guidelines for Lithium-ion (Li-ion) Batteries NASA/TM-2010-216727/Volume II, NESC-RP-08-75
44. Walter PL, Pyroshock Explained, Technical Note TN-23, PCB Piezoelectronics, Inc.
45. http://www.dodbuzz.com/2013/05/21/navy-developing-hybrid-electric-amphibious-assault-ships/. Accessed on June 25, 2013
46. Doerry N, Fireman H (2006) Designing All Electric Ships. Ninth International Marine Design Conference. Ann Arbor, Michigan. May 16–19, 2006
47. http://www.nytimes.com/2012/08/28/science/earth/cargo-ship-designers-turn-to-wind-to-cut-cost-and-emissions.html?_r=0. Accessed on June 25, 2013

48. Noble P (2009) Past and Present Use of Electric Ships in the Energy Industry. IEEE Electric Ship Technologies Symposium. Accessed on June 25, 2013
49. http://news.panasonic.net/archives/2012/0629_11250.html. Accessed on June 25, 2013
50. http://www.gereports.com/the-pulse-of-the-sea/. Accessed on June 25, 2013
51. Puglia F, Cohen S, Hall J, Yevoli V (2005) Very Large Lithium Ion Cell and Battery Designs. The 5th International Advanced Automotive & Ultracapacitor Conference (AABC-05) Proceedings. Honolulu, Hawaii
52. Rausand M, Høyland A (2004) System Reliability Theory: Models, Statistical Methods, and Applications, 2nd Edition, John Wiley & Sons, Hoboken, NJ

Chapter 5
Lithium-ion Cell and Battery Safety

Seth Cohen, Malgorzata K. Gulbinska and Frank J. Puglia

Abstract Lithium-ion cell and battery safety has recently emerged as a major topic of research and development work. This chapter will focus on identifying the leading safety hazards in a lithium-ion cell and battery, defining the currently taken pathways to address these hazards and highlighting the possible future safety solutions. Statistically, lithium-ion cells and batteries are very reliable. They are present in nearly all cell phones and portable electronic devices today. Yet, as more energy is being packed into these small spaces, it has become imperative for engineers to address safety at all levels of design from the choice of active materials to full system topography.

5.1 Battery Safety: An Introduction and Critical Definitions

Since their inception on the market in the early 1990s, Li-ion batteries have rapidly become one of the most attractive technology-enabling energy storage devices, partially due to their continued ability to push the envelope with respect to higher energy and power demands. The same qualities that make these batteries attractive for use in these applications, correspondingly contribute to the potential safety hazards.

S. Cohen (✉) · M.K. Gulbinska · F.J. Puglia
Yardney Technical Products, Inc., 2000 South County Trail, East Greenwich, RI 02818-1530, USA
e-mail: scohen@yardney.com; Seth.H.Cohen@gmail.com

M.K. Gulbinska
e-mail: mgulbinska@gmail.com

F.J. Puglia
e-mail: fjpuglia@yardney.com; fjpuglia@gmail.com

M.K. Gulbinska (ed.), *Lithium-ion Battery Materials and Engineering*,
Green Energy and Technology, DOI 10.1007/978-1-4471-6548-4_5

Currently, lithium-ion batteries provide power for platforms ranging from portable electronics and power tools to cars. Electric vehicles' progress is marked by the developments in hybrid electric vehicles (HEV) such as Nissan LEAF® and Chevy Volt and plug-in electric vehicles (PHEV) like Tesla Model S. In the ongoing efforts to maximize run-time, range, and power of these applications, materials with higher reactivity are being developed, which increases the impact that the potential catastrophic failures could have in the field.

Like any other energy-dense power source, lithium-ion batteries contain highly energetic materials, in this case: cathode and anode. The electrodes are in contact with a flammable electrolyte based on organic carbonates. Under normal use conditions, lithium-ion batteries deliver the stored energy safely and at the requested rate. However, when subjected to abusive conditions, lithium-ion cells can undergo the *thermal runaway* (Text Box 1) and release a significant amount of energy in the form of heat, which in turn can lead to fire and explosion.

Text Box 1 Thermal runaway in lithium-ion cells [1, 2]

Thermal runaway of lithium-ion batteries is often described as a three-step, self-propagating process:

(1) In the first (rate-limiting) step, immediately following the cell's abuse, anodic reactions start at about 80–90 °C. As temperature rises above 120 °C, decomposition of the SEI layer accelerates, leading to reduction of the electrolyte at the lithiated graphitic negative electrode.

(2) In the second step, exothermic reactions at the positive electrode start as the temperature rises above 140 °C. Typically, oxygen rapidly evolves from the cathode material at this stage.

(3) The cathode decomposes and the electrolyte becomes exothermically oxidized at the onset temperature of >180 °C, yielding at high rate temperature rise (≥100 °C per minute).

Thermal runaway and heat evolution in lithium-ion cells depend on the cell's state of charge (the higher the charged voltage the lower the process onset temperature) and on the cycling history of the cell, the applied load and the type of the cell.

The diagram in Fig. 5.1 illustrates the main causes of failure in lithium-ion batteries, followed by the exothermic reactions that typically occur after abuse and the hazards that result from the battery undergoing a violent energy-releasing event. Some of the terms in Fig. 5.1 are self-explanatory (e.g., mechanical or thermal abuse) while others, such as the electrical abuse conditions, require more detailed description. The electrical abuse conditions, whose discussion follows below, include: overcharge, over-discharge, or high (i.e., higher than the battery power rating) pulse power exposure.

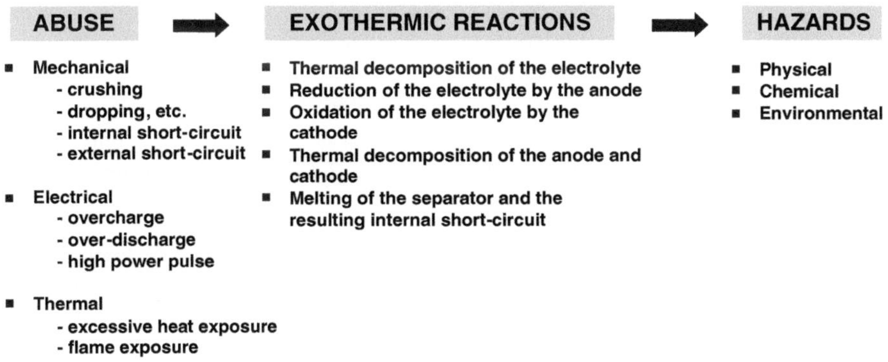

Fig. 5.1 Top-level diagram of the failure causes, with the resulting reactions and hazards in lithium-ion batteries [3]

The *overcharge*-related behavior of the lithium-ion chemistry is best explained starting from the overall cell equation, shown in a generalized form in Eq. 5.1.

$$LiMO_2 + \frac{x}{y}C \xrightarrow{CHARGE} Li_{1-x}MO_2 + \frac{x}{y}Li_yC \tag{5.1}$$

During the cell's "normal" charge, lithium ions migrate from lithium ion mixed metal oxide (commonly $LiMO_2$, where M = Co, Ni, and/or Mn, etc.) cathode to anode (typically, carbon). Considering the charge balance in cells' materials as the charge process progresses, the cathode becomes more oxidative and the anode becomes more reduced in character and both electrodes are brought to higher energy state, while the cell voltage increases. If for some reason (e.g., incorrect packaging or handling), the cell's voltage exceeds the "safe" limits (these are different for various electrode chemistries; discussed in Chap. 1), the overcharge process begins, which is always accompanied by heat evolution.

The overcharge mechanism generally accepted in the literature is similar to that proposed by Ohsaki et al. [4] for prismatic $LiCoO_2$ cells:

(1) Migration of lithium ions between electrodes continues as the cell voltage increases. Gas evolution and temperature rise in the outer casing are low.
(2) As overcharge approaches 100 %, positive electrode resistance increases, resulting in heat generation due to the Joule effect and the exothermic decomposition of the organic electrolyte solution. The electrolyte is decomposed by Co^{4+} ions whilst the remaining $LiCoO_2$ undergoes structural transformation.
(3) Rapid temperature rise occurs due to reaction between the lithium-depleted, highly energetic cathode and the electrolyte. The reaction accelerates as the cell's internal temperature rises above 60 °C and CO_2 gas is generated from oxidative decomposition of organic electrolyte.
(4) As the temperature approaches 130–135 °C, the cell may behave according to one of the following pathways:

(a) If the polyethylene separator shutdown occurs, which is most likely in the event of a low-rate (<1C) overcharge, the temperature would decrease with no rupturing.
(b) Higher rates of overcharge could lead to thermal runaway and eventual cell rupture/venting as a highly exothermic reaction between the highly reductive negative electrode (possibly, containing some metallic lithium deposits[1]) and the electrolyte occurs [2].

As stated before, the energy release from materials in batteries can be benign (normal operation), mild (e.g., low-rate overcharge with sufficient heat dissipation), or violent (thermal runaway). If a battery undergoes a violent energy release, the following hazardous conditions arise:

(1) Physical hazards, such as a simple rupture of battery case.
(2) Chemical hazards from leakage or venting of corrosive or toxic materials in the battery.
(3) Environmental hazards arise from the reactive and flammable nature of lithium and/or leakage of toxic materials from batteries that are improperly disposed [2].

Both chemical and physical hazards can cause further equipment damage due to breakage or corrosion of electrical/electronic components [2].

The *over-discharge* of a lithium-ion battery is a relatively uncommon condition but is nevertheless possible in specific conditions. One well-documented (and very interesting) instance of the over-discharge occurred in several cells of the lithium-ion battery ($LiCoO_2$ cathode vs. graphitic anode) used on Japanese interplanetary spacecraft HAYABUSA and will be quoted here to explain the over-discharge mechanism and the resulting battery damage symptoms [5]. In HAYABUSA's case, a loss of communication with the spacecraft resulted in four out of eleven cells being over-discharged by an auxiliary circuit that kept draining power from the battery, while several of the cells were insufficiently recharged. The forced over-discharge "extended" the standard discharge that would normally end at 3.0 V and the affected cells were drained completely to 0 V. An attempt at recharge of the damaged cells resulted in a quick rise of voltage to 0.3 V, where eventually a prolonged plateau was observed. When the recharge current was discontinued, an immediate voltage drop to 0 V was observed. The following explanation was formulated: when the lithium-ion cell was discharged to 0 V, part of the copper current collector was dissolved in electrolyte solution. Upon the attempted recharge, the copper was plated onto the graphitic anode, forming an internal short-circuit that kept draining the cell in the absence of current or if the low-rate recharge current was applied. The high-rate current resulted in an abrupt voltage increase and resistive heating of the cell. Since the robotic spacecraft needed the battery power for the safe return to Earth and all the cells (both damaged and undamaged) were

[1] Under normal, non-abusive conditions, there is no metallic lithium in a lithium-ion cell.

connected in series, a very low current was used to slowly recharge the undamaged cells to their normal top voltage of 4.1 V, while in the damaged cells the continued passing of low-rate electrical current resulted in the eventual plating of copper at the positive electrode due to the negative potential of the cells (−0.3 V). Ground battery testing and over-discharge simulations were used to ensure the safest operation of the damaged battery.

The *high pulse power* exposure is the third type of electrical abuse that can potentially be experienced by lithium-ion batteries and lead to their catastrophic failure. This type of abuse is the most difficult to recognize and quantify since the term "high power" means something very different for various battery designs that are optimized for various energy and power ratings. More detailed description of the lithium-ion technology optimization can be found in Chap. 3 of this book. Below, the main reasons for safety hazards associated with the high power exposure are discussed using an HEV battery as an example.

For pulse power operation typical of HEV applications, heating (also known as self-heating) is a major issue. Ohmic heating dominates other heating mechanisms over a wide range of temperatures and battery's states of charge (SOC) typically seen in HEVs. Self-heating of the battery, if unchecked, increases the likelihood of thermal runaway and therefore, the majority of the evolved heat must be dissipated, often using forced air convection. More heat is generated in a battery that is aggressively cycled and/or utilized at lower temperatures (e.g., less than 25 °C). Slow lithium-ion diffusion within electrode materials (especially the carbon anode) was found to be the culprit that limited high rate performance, especially at subzero temperatures. Slow diffusion causes lithium-ions to accumulate near the surface leading to a concentration gradient within active material particles (surface vs. bulk). This can pose further safety hazards such as the possibility of side reactions on the negative electrode, even possibly leading to the metallic lithium deposition during high rate charge pulses. Further, lithium-ions concentration gradient can lead to either oversaturation or depletion of lithium-ions from the surface layers or either cathode or anode, depending on the exact character of the pulse (charge or discharge). In cathode materials, the depletion of the surface layers leads to structural changes and deterioration of cathode's performance [6].

Since different active materials used in lithium-ion cells have very different responses to the cell's abuse, the materials selection is crucial from the battery safety and performance point of view. The battery design process usually starts from the selection of cell chemistry and then the safety of the system is analytically determined from the summary contribution of the individual components. Chapter 1 presented an introduction to the constituents of the lithium-ion cell: the anode, cathode, separator, and electrolyte. This chapter will discuss safety impact of these components, as well as the internal cell protection devices. A discussion of the selected options of the cell design and packaging (e.g., cell aspect ratio) in terms of thermal and safety properties will be also included. The chapter will conclude with examples of single cell and cell pack safety testing.

5.1.1 Analyses Methods: From Materials, Through Components, to Cells

Concurrently with analyzing novel materials for increased capacity and rate capability, safety testing of these cell components must be done, first on the raw material and then, in the actual lithium-ion cells. There are several experimental and analytical methods to help assess the reliability, reactivity, and sensitivity of the cell materials to thermal runaway and their stability. Two of the more common methods of identifying effects that thermal abuse will have on the components are differential scanning calorimetry (DSC) and accelerating rate calorimetry (ARC).

The DSC method measures the amount of energy (heat) absorbed or released by a sample as it is heated or cooled relative to a reference (e.g., copper) [7]. DSC allows for direct measurements of the thermal response over a broad temperature range. Small samples are typically placed in an open aluminum or hermetically sealed stainless pan and heated at a fixed rate of 5 °C · min^{-1} or 10 °C · min^{-1}. The DSC is a useful tool in analyzing the temperatures at which the material undergoes physical and/or chemical transformations, for instance, temperature at which phase transitions or chemical reactions within the material occur. Depending on whether the reaction is endothermic or exothermic or what type of phase change is occurring, the DSC instrument will need to adjust the flow of heat into or out of the sample holder. Output from the DSC, as can be seen in Fig. 5.2, is presented in the form of a curve (or a table) that represents heat flow at the range of temperatures evaluated.

DSC measurements are good "first estimates" of the thermal effects related to decomposition of materials in various SOCs, with or without electrolyte. The main limitation of the DSC method is that milligram amounts of material (especially important with electrolyte, which is delivered by micropipette) are used during measurements. This means that relatively small experimental variations can result in considerable differences in the heat flow. It is noteworthy that onset temperatures for exo- and endothermal effects measured by DSC are more reproducible than the absolute heat values and, as such, more practical for interpretation. Understanding the DSC method's limitations allows for the accurate estimation of the trends in materials thermal performance and the results are later confirmed by further studies, in full cells of various sizes and designs.

ARC testing provides more precise temperature control than DSC, under adiabatic conditions. Therefore, the onset of self-heating due to chemical reactions can be detected with greater sensitivity [8]. ARCs operate by heating the chamber and sample by a specific discrete temperature, then hold that temperature for a preset amount of time while monitoring the temperature of the sample. If the sample's temperature is increasing at a rate that does not exceed certain threshold, the ARC will increase the temperature of the chamber by another discrete temperature step. If the rate of rise of the sample exceeds the threshold value, the ARC will attempt to match the sample's heating rate and temperature to maintain adiabatic conditions. Typical output from the ARC is given in the form of a table or curve representing the rate of temperature rise of the sample over the range of temperatures

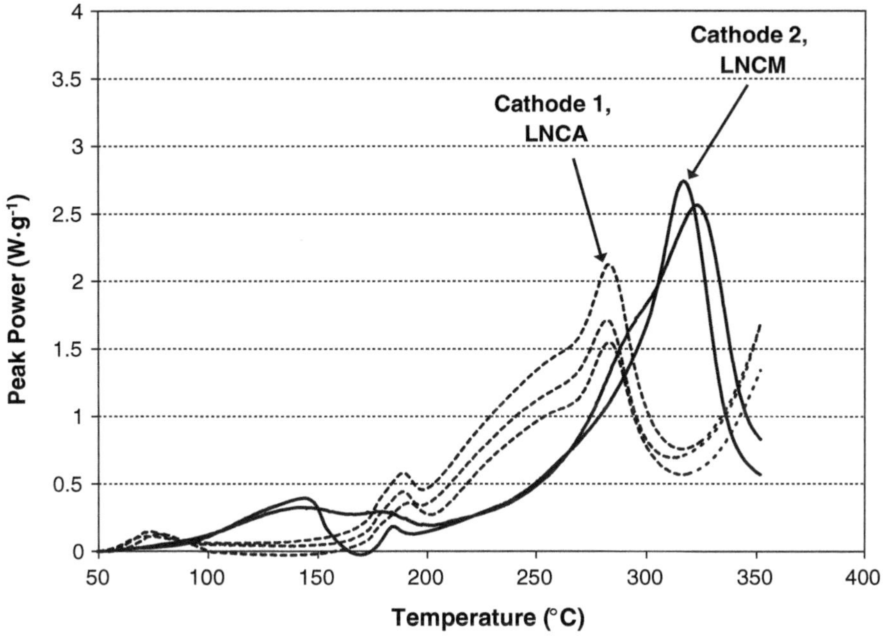

Fig. 5.2 DSC results for two different cathode chemistries: "*Cathode 1*": $LiNi_{0.8}Co_{1.5}Al_{0.5}O_2$ (LNCA), and "*Cathode 2*", $LiNi_{0.33}Mn_{0.33}Co_{0.33}O_2$ (LNCM), run at 5 °C · min^{-1}; samples were fully charged, washed in dimethyl carbonate (DMC), and dried overnight; the active material was then scraped and run in the DSC

evaluated. In addition to the adiabatic measurement conditions, another advantage of the ARC is the allowable size and the complexity of the sample; full prismatic cells with capacities as large as 7 Ah can be analyzed by this method. Of course, obtaining thermal behavior data from the cells' components and/or raw materials prior to the ARC testing of the full, large cells is highly advisable for the best interpretation of the results.

5.1.1.1 Cathodes

The most prominent cathode materials used in today's commercial lithium-ion cells were discussed in Chap. 1 in terms of their syntheses, structure, physical, and chemical properties as well as their cell performance. Cathodes are known for their reactivity, especially when combined with electrolyte solutions and at elevated temperatures, which significantly impacts safety at the cell level. Cathodes are highly electronegative materials with high potential versus lithium metal. The cathode needs to be a strong oxidizing agent and has the ability to accept and store lithium cations from the electrolyte. Therefore, these compounds typically have high oxygen content within their molecules.

Table 5.1 Selected positive electrode materials and their abbreviations

Cathode material	Abbreviation
$LiCoO_2$	LCO
$LiNiO_2$	LNO
$LiNi_{0.8}Co_{0.2}O_2$	LNCO
$LiNi_{0.375}Co_{0.25}Mn_{0.375}O_2$	LNCM
$LiMn_2O_4$ (spinel)	LMS
$LiFePO_4$	LFP

Based on their structure, cathodes can be separated into three basic families; layered oxides, spinels, and olivine phosphates. The three families differ greatly in their cell performance as well as in thermal properties. Layered oxide cathodes typically have higher specific capacities than spinels or phosphates. As a consequence of their structure, layered oxide materials are more energetic upon thermal runaway events due to the ease with which oxygen is liberated from their framework upon decomposition.

Phosphates have the advantage of the strong phosphorus-oxygen bonds in the phosphate (PO_4^{3-}) anion, making it more difficult for oxygen to be freed than from other cathodes. Therefore, higher temperatures are required for thermal runaway to commence in phosphate cathodes-based cells. The danger of oxygen liberation is that a high temperature failure will continue to be fueled and provide for an explosive environment with the highly flammable electrolyte also present in the cell.

Due to their reactivity with electrolyte solutions, especially in the charged state, the most useful thermal behavior testing (e.g., DSC) of the candidate cathode materials occurs in the presence of electrolyte. Below, the results of a comprehensive study done by MacNeil et al. [9] have been used in preparation of graphs shown in Figs. 5.4, 5.5, 5.6, and 5.7, with several cathode/electrolyte system behavior trends identified and discussed; common cathode name abbreviations used in the discussion are listed in Table 5.1. Similar studies are commonly done by the lithium-ion battery manufacturers on various cathode and/or electrolyte materials candidates and are usually treated as a basic introduction to the ARC testing and other safety-related experiments on larger cells (overcharge, nail penetration tests, etc.).

Before most of the DSC studies, including the one discussed below, the materials are charged to a desired voltage in the electrochemical coin cells (the name refers to their approximate size and shape; several models exist) against Li/Li^+ anode (half-cells). After the disassembly of the cell, the cathode is taken out and washed in an organic solvent to remove traces of lithium hexafluorophosphate ($LiPF_6$), dried, and loaded into the DSC pan.

In the study done by MacNeil et al. [9], all of the major cathode types (layered oxides, spinel, and a phosphate) were included. The maximum reactivity of electrodes was ensured by precharging the materials to several high voltage levels and by adding electrolyte solution into the DSC vessel. Several general trends in cathodes' cell performance and thermal behavior were observed. Figure 5.4 illustrates the precharging results and gives an indication on the capacity delivered by various cathodes at the given voltage values. The operating voltage range up to about 4.2 V is currently the most desired for the lithium-ion battery cathodes, because in the 4.2–4.4 V, shortened cycle life is observed in most materials and above 4.4 V, electrolyte oxidation occurs on highly charged cathodes. Therefore, the cathodes that can deliver large portion of their capacity under 4.2 V are the most sought after.

The first cathode material that was found suitable for the commercial rechargeable lithium-ion cells, $LiCoO_2$, is still in use despite several shortcomings. $LiCoO_2$ delivers modest specific capacity but, given its high operating voltage versus lithium, has a high energy density (Table 1.2, Chap. 1). Furthermore, $LiCoO_2$ has high skeletal density and thus can be used in high energy density batteries. Due to these reasons, $LiCoO_2$ is still widely used in the market today. There are several drawbacks to this material, however. Lithium cobalt oxide can only be discharged at relatively low currents. Due to the large internal resistance, at high discharge currents, the cells are subjected to elevated temperatures and the safety of the cells is compromised. $LiCoO_2$ has a short cycle life as internal resistance also increases at a rapid rate with cycling, leading to even more heat generation.

The attractiveness of lithiated nickel oxide as a potential replacement for lithium cobalt oxide can be clearly seen in Fig. 5.4, where $LiNiO_2$ delivers 194 mAh•g^{-1} at 4.2 V versus 140 mAh•g^{-1} from $LiCoO_2$ obtained at the same voltage. Lithium nickel oxide, $LiNiO_2$, is isostructural with $LiCoO_2$ and has long been proposed as an alternative cathode, given the higher energy density and relatively low cost of nickel compared to cobalt [10]. There are several reasons that this material has not been adopted into the market. As pointed out in Chap. 1, it is quite difficult to synthesize this material in a defect free, stoichiometric $LiNiO_2$. These low lithium content compounds are unstable due to the high equilibrium oxygen partial pressure, so the $LiNiO_2$-based cells are inherently unstable. As seen in Fig. 5.5, severely delithiated (overcharged) $LiNiO_2$ exhibits one of the most vigorous exothermic reactions during breakdown compared to other cathode materials. The thermal behavior difference between $LiNiO_2$, $LiCoO_2$, and $LiNi_{0.8}Co_{0.2}O_2$ is less pronounced in the total amount of the evolved heat (J g^{-1}) on overcharge (Fig. 5.6) than in the rapidity of heat evolution (Fig. 5.5) evidenced by the peak power (W · g^{-1}). Considering that both the total evolved heat and the peak power should be as low as possible and the onset temperature (Fig. 5.7) should be as high as possible for a cathode material to be attractive from the safety point of view, $LiNi_{0.375}Co_{0.25}Mn_{0.375}O_2$ and the related $LiNi_{1-x}$-yCoxMnyO$_2$ materials emerge as the safer, yet reasonably energetic cathode materials. Capacity of 158 mAh · g^{-1} for $LiNi_{0.375}Co_{0.25}Mn_{0.375}O_2$ at the reasonably high mid-point voltage (~3.8 V vs. Li/Li^+ anode), combined with the consistently high onset temperatures (>225 °C between 4.0 and 4.6 V; Fig. 5.7) and relatively low cost of manganese compared to cobalt and nickel explain why there is an interest in LNCM materials for various applications, including the electric vehicles.

The structure, cell performance, and limitations of lithium manganese spinel ($LiMn_2O_4$) were discussed in detail in Chap. 1 of this book. As a direct consequence of the stable structure, spinels hold oxygen more strongly than the layered oxides during overcharge and thermal abuse. The resulting peak power and total evolved heat are low in spinels, as seen in Figs. 5.5 and 5.6, while the onset temperature

(Fig. 5.7) is above 250 °C at 4.2 V. The thermal behavior of $LiMn_2O_4$ charged to 4.2, 4.4, and 4.6 V does not change significantly, which is due to the fact that very little lithium remains in the $LiMn_2O_4$ structure at these voltages and thus the electrodes are almost identical. Although $LiMn_2O_4$ at 4.2, 4.4, and 4.6 V experiences thermal instability in the 220–200 °C range, very little heat is released (Figs. 5.5 and 5.6).

$LiFePO_4$ is another cathode material, next to lithium manganese spinel, that releases most or all of lithium-ions at relatively low voltage. Specifically, the mid-point voltage for $LiFePO_4$ is at 3.45 V (in a half-cell), while the charge/discharge profile is very flat, leaving no lithium ions in the LFP structure when the 3.8 V limit is reached. Although the theoretical capacity of LFP is low (169 mAh g^{-1}, compared to about 274 mAh g^{-1} in both $LiCoO_2$ and $LiNiO_2$), preparative methods have been developed such that nearly all the lithium (160 mAh $\cdot$ g^{-1}) in the structure can be extracted and reinserted without significant capacity or life losses. The 95 % of lithium ions removed reversibly from LFP can be compared to only about 51 % of lithium ions extracted from $LiCoO_2$ and 71 % from $LiNiO_2$. Compared to the Li_xCoO_2 in a similar delithiation stage (170 mAh g^{-1} extracted at 4.4 V from LCO), LFP at 3.8 V after the removal of 160 mAh g^{-1} shows a very small exothermal peak (Fig. 5.6) and the low peak power (Fig. 5.5). The trade-off for the good structural stability of $LiFePO_4$ and the minimal tendency to release oxygen (due to the strong P–O bonds) under abusive conditions is seen in lower electrochemical energy per unit mass and volume. The low specific energy and energy density of phosphates compared to layered oxides is easily understood looking at the $LMPO_4$ versus LMO_2 formulas (M = metal; e.g., Fe or Co, Ni) where in phosphates one lithium atom is counted per the entire PO_4^{3-} moiety (95 g mol^{-1}) compared to one lithium per two oxygen atoms (32 g mol^{-1}) in layered oxides.

Despite the great small-scale (DSC) results in a highly charged stage and with electrolyte, fires, and explosions have been reported even for the “safe” $LiFePO_4$-based batteries in electric vehicles for a variety of reasons [11]. This is another reminder that lithium-ion battery materials are highly energetic in nature and require comprehensive safety measures that are not limited to the cell chemistry selection. The remaining part of this chapter discusses the lithium-ion technology safety in terms of anode, electrolyte, and separator contributions, followed by the cell-level protection devices and battery thermal modeling. Finally, the electronic battery management system (BMS) considerations are reviewed.

5.1.1.2 Anodes

On the anode, the surface of carbon electrode is covered with the products of the electrolyte reduction which form a protective solid electrolyte interface (SEI) [12–15]. SEI layer on the anode’s surface controls the safety behavior of lithium-ion cells to a great extent. This belief is supported by voluminous research done predominantly on carbon-based anodes, such as graphites and cokes (Chap. 1), cycled with solutions of lithium-ion salts in organic carbonates. With the emerging anodes, such as silicon, silicon-carbon composites, and Sn–Co–C compounds, the SEI formation has not been yet fully explored [16] but there is strong experimental evidence that SEI is indeed formed on these anodes. The stipulation that SEI formed on noncarbonaceous anodes would also affect the cells' safety is not particularly far-fetched.

Text Box 2 SEI formation in lithium-ion cells [17]

In the $LiPF_6$ salt and organic carbonates containing electrolytes, the SEI film is a passivating surface film on the anode (e.g., a graphite particle) consisting of Li-alkyl-carbonates, polymeric carbonates, and lithium fluoride and is formed on the surface of particles via either the direct decomposition of the solvents to form a precipitate layer, or the co-intercalation of solvated lithium ions and their decomposition beneath the surface in the initial stage of lithium intercalation (exfoliation), during the first few cycles.

According to the current research, the first stage of the thermal runaway is initiated at the anode by the breakdown of the thin protecting SEI layer. Specifically, with carbon anodes and the $LiPF_6$ salt dissolved in organic carbonates, thermal runaway starts at approximately 80 °C, which is a rather low temperature that can be reached by exposing the cell to elevated ambient temperatures, overcharging the cell, or exposures to high charge or discharge currents. As the protection of the SEI layer breaks down, the electrolyte begins to exothermically react with the anode which prompts further increase of the cell's temperature and, if unchecked, is followed by: flammable hydrocarbon gases generation, internal cell pressure increase, separator meltdown (120–140 °C), and finally, the highly exothermic, oxygen-releasing cathode decomposition. The end result is the combustion of the highly pressurized flammable gases that have been generated within the cell.

Lithium bis(oxalato)borate (LiBOB) salt is believed to form a robust SEI on lithiated carbon surface, preventing any exothermic process until 170 °C as compared to an onset temperature of 80 °C with $LiPF_6$–ethylene carbonate (EC)–diethyl carbonate (DEC) [3]. Better SEI formation properties and higher decomposition temperature of LiBOB as well as lack of propensity to form trace amounts of HF in the presence of moisture, render LiBOB the safer salt candidate than $LiPF_6$ but the hexafluorophosphate is still viable due to better solubility in the typical electrolyte solvents and higher conductivity than LiBOB (Chap. 1, Tables 1.8 and 1.9).

5.1.1.3 Electrolytes

As stated before, the conventional lithium-ion electrolytes are generally comprised of lithium-based salts in mixtures of one or more organic solvents. More often than not these organic solvents are linear or cyclic alkyl carbonates (Chap. 1, Table 1.8) that are very volatile and, under certain conditions, are extremely

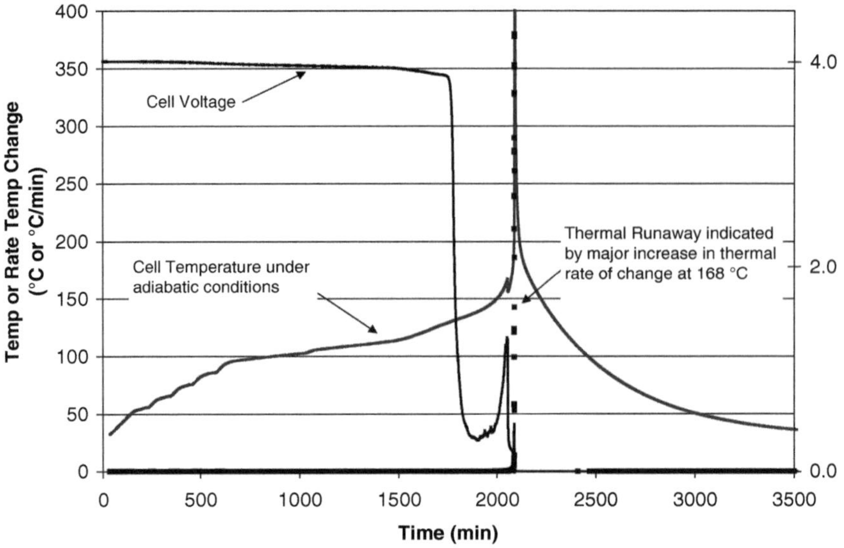

Fig. 5.3 ARC results plot of a fully charged cell

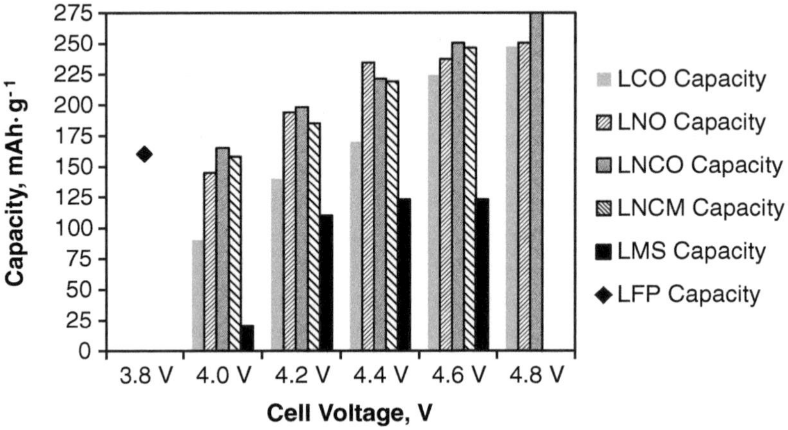

Fig. 5.4 Cathode materials results summary: capacity versus voltage

flammable. Furthermore, when exposed to elevated temperatures and voltages, the electrolytes can react undesirably with the active electrodes (Text box 1, Figs. 5.2, 5.3, 5.4, 5.5, 5.6, and 5.7). These reactions can generate excessive heat and gases causing extreme pressure to build up within the cell. Some of the gases evolved during decomposition are flammable hydrocarbons such as hydrogen (H_2), methane (CH_4), or ethane (C_2H_6), etc. If sufficient pressure is generated, cell venting occurs, forming highly flammable, explosive mixture of fuel and air. Increased vapor pressure can vent cells at temperatures as low as 150 °C [2].

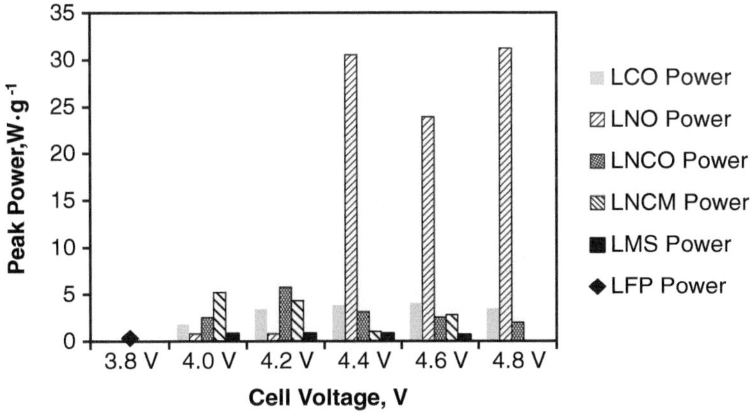

Fig. 5.5 Cathode materials results summary: DSC peak power values

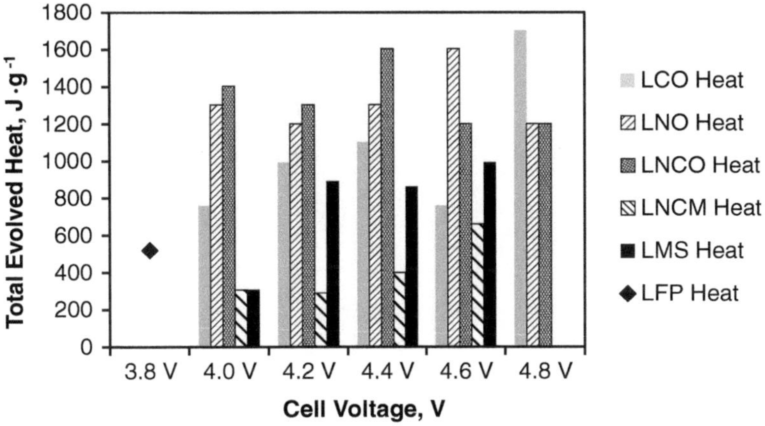

Fig. 5.6 Cathode materials results summary: DSC total evolved heat

Compounding these hazards is the fact that during battery failures, especially in high voltage applications, it is not uncommon for sparks to be present that can ignite the mixture. Sparks are not a prerequisite for the ignition to occur as the combined heat of the runaway reactions can often be sufficient to cause the vented gases to spontaneously ignite, specifically if the autoignition temperature of the mixture is exceeded.

Analyzing the mixtures of gases and solid products generated during various cell failure scenarios (e.g., overcharge and/or heat exposure, etc.) gives insights into the mechanism of the thermal runaway. Gas chromatography-mass spectrometry (GC-MS) is commonly used to separate (GC) and identify (MS) gases generated during the controlled safety event. Often, X-ray diffraction (XRD) is used on solid breakdown products to establish or confirm their structure and X-ray

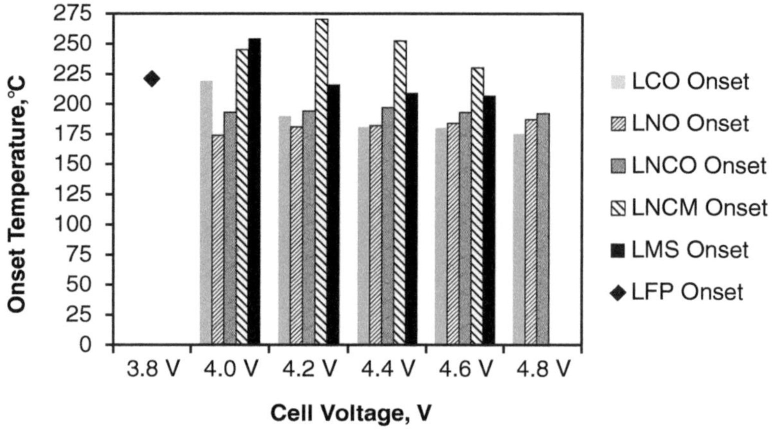

Fig. 5.7 Cathode materials results summary: DSC onset temperatures

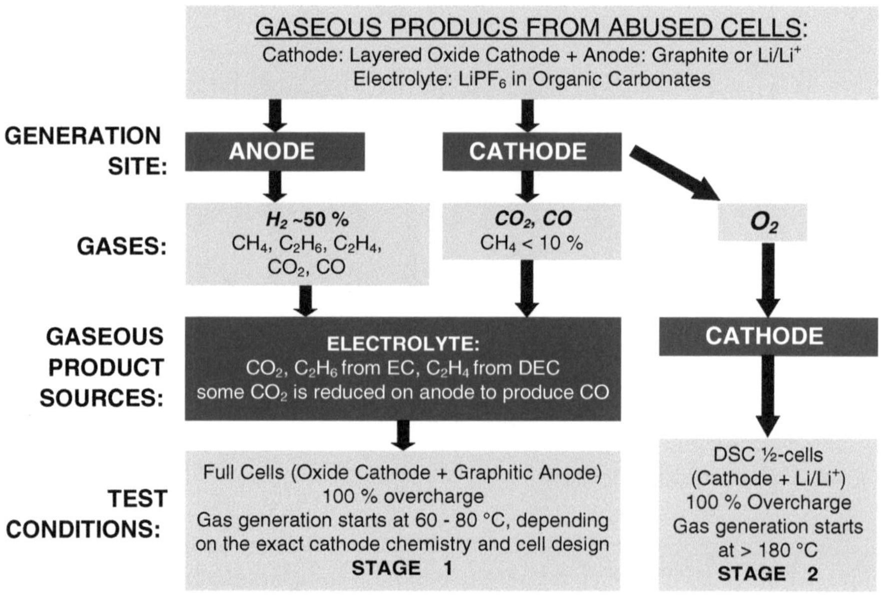

Fig. 5.8 Gaseous products of electrolyte decomposition [1, 2, 4]

photoelectron spectroscopy (XPS) is used to identify the concentration and chemical state of elements near the sample's (electrode's) surface.

A diagram of gaseous products generated inside a lithium-ion cell under abusive conditions is shown in Fig. 5.8. The results of several studies done in varying conditions were used to prepare the diagram; the data (such as the exact onset temperature or the exact contents of gases) are expected to somewhat differ in various cell chemistries (i.e., cathode, anode, salt, and solvents) but the main trends and major products will most likely stay consistent. In general, gaseous products result mainly from electrolyte decomposition that may be additionally catalyzed by $LiPF_6$ reaction products [18], which are mostly nongaseous (solids or liquids) and end up being deposited on the surfaces of electrodes. A thick layer of solid LiF deposits and inorganic and organophosphate compounds were found on graphitic anodes from abused cells. The cathode surface films contained LiF, Li_2O, $Li_xPO_yF_z$, and possibly small amounts of alkyl fluorophosphates species, also coming from the breakdown of electrolyte solution [1].

Flame-retarding electrolytes. Many efforts have been focused on the development of flame retardant electrolytes for lithium-ion batteries. Most of the work concentrated on the incorporation of phosphorus-containing flame-retarding additives to standard $LiPF_6$ based electrolytes [20–27]. Some limited work was done on incorporation of different salts (other than $LiPF_6$) or salt additive combinations with flame-retarding additives [19]. The rationale behind this direction of work is that phosphorous- (and halogen-) containing compounds react with hydrogen, which is one of the main combustible electrolyte decomposition products (Fig. 5.8). However, incorporation of the phosphorus-containing flame retardants typically results in the increased cell performance losses, many of which have been attributed to disruption of the SEI formation process by the presence of the flame retardant. The disruptive effects of flame retardants are compounded by the fact that large quantities (10–20 wt% of electrolyte solution) of these compounds are needed in the cell to have the sufficient effect on safety of the cell.

The past efforts on development of organophosphorus-based flame-retarding compounds have focused on trialkylphosphates such as trimethyl phosphate (TMP) [20], triethyl phosphate (TEP) [21], tris(2,2,2-trifluoroethyl) phosphate (TFP) [22], and bis(2,2,2-trifluoroethyl)methyl phosphate (BMP) [23], cyclophosphazenes such as hexamethoxycyclophosphazene (HMPN) [4], phosphoramides including hexamethylphosphoramide (HMPA) [24], and phosphonates including dimethylmethyl phosphonate (DMMP) [25–27]. However, none of these materials have functioned ideally. Trialkylphosphates and phosphoramides have problems associated with reduction on the battery anode when phosphate concentration is high enough for effective flame retardance. HMPN has good electrochemical stability toward the electrodes but the high concentrations required for reduction of flammability result in increased viscosity and lower charge/discharge rates. The best flame-retarding additives currently reported are TFP and DMMP.

SEI formation-boosting additives. There has been a significant interest in the development of electrolyte additives to enhance the stability of $LiPF_6$-based electrolytes for lithium-ion batteries [28]. One class of additives, which includes vinylene carbonate (VC) [29] and LiBOB [30, 31], react on the surface of the anode to generate a more stable SEI. Another class of additives includes N,N-dimethylacetamide (DMAc). DMAc reduces the reactivity of $LiPF_6$ and inhibits the reactions between the electrolyte and the electrode materials [32–34]. Stabilizing the anode SEI and inhibiting the detrimental reactions of the electrolyte with the surface of the electrode materials will extend calendar life and enhance the thermal stability of lithium-ion batteries.

Research and development of electrolyte additives that react with the surface of the cathode (mainly to improve stability at high voltages and/or high state of charge) have primarily focused on overcharge protection via either redox shuttles or shutdown polymerization reactions, with some work being done on additives designed to generate thin cathode surface films similar in nature to the anode SEI has been limited [35].

Redox shuttle additives. The (idealized) mechanism of operation of the redox shuttle electrolyte additives involves electrochemical oxidation at the positive electrode at potentials that are slightly higher than the typical charging voltage. The oxidized additive molecules then diffuse to the negative electrode, where they become reduced back to the starting neutral molecules, which then shuttle back to the positive electrode. Thus, redox shuttles shunt the excess charge injected into the cell during overcharge [3]. There are several conditions that the potential redox shuttle materials should fulfill:

(1) Both the oxidized and reduced forms of the redox shuttle molecule must be inert toward cell components and possess thermal stability over the temperature range of the cell's operation.
(2) The solubility, mobility, and diffusion coefficient of both the oxidized and reduced redox shuttle species in the battery electrolyte should be high.
(3) The oxidation potential of the redox shuttle molecules must be lower than the decomposition potential of the electrolyte solvents but slightly higher than the overcharge cutoff voltage.
(4) The shuttle must be electrochemically reversible and must not cause any side reactions.
(5) The reversibility of the couple should last for the entire cycle life of the cell.

There are several barriers on the road to successful implementation of redox shuttles in lithium-ion cells; e.g., several suitable redox couples work only at high charging voltages, and this means that they actually do not respond to heat generation in batteries. Very few stable redox shuttles for high voltage cathodes have been reported so far, presumably due to their high reactivity. Eventually, the majority of redox shuttle additives fail, presumably due to decomposition of their radical cation forms. Increased electron deficiency makes radical cations more susceptible to nucleophilic attack, which may result in reactions with electrolyte components.

Polymer and polymer-gel electrolytes. These electrolytes represent another approach to achieving better safety characteristics in lithium-ion batteries. Several polymer and polymer-gel electrolytes have been investigated and have demonstrated improved flame resistance [36]. However, the improvements were accompanied with problems including significantly lower conductivity, especially at low temperature. Solid state electrolytes are another, safer alternative to the organic carbonates-$LiPF_6$ solutions. Due to the inherent low conductivity of solid electrolytes, they are not practical for many applications where high power and low temperature performance are required. Ionic liquids are nonvolatile, nonflammable, highly conductive, and environmentally benign and operating within wide temperature range. However, the cost of ionic liquids is prohibitively high and therefore, their mass commercialization.

5.1.1.4 Separators

Although electrochemically inert, separators perform a critical function in the lithium-ion cell. The main role for separators is to prevent direct electrical contact between the cathode and anode, while maintaining the least tortuous path for Li^+ ion migration between the two. Under normal discharge conditions, the negatively charged electrons move from the anode, through the negative electrode's current-collector, to the outside world where they perform work. The electrons continue their journey through wiring, back into the cell, through the positive electrode's current-collector and into the cathode. Congruently, the positively charged lithium ions, created as the electrons leave, move from the anode, through the separator and into the cathode where they rejoin with the electrons. Accordingly, there is no reason for either physical or electrical contact between the anode and cathode at any point.

Most of the commercially available separators consist of single or multilayer porous polyolefin sheets, with polyethylene (PE) and polypropylene (PP) being the most commonly used polyolefins. Polyethylene terephthalate (PET) and polyvinylidene fluoride (PVDF) have also been used in separators, although not extensively. The main advantage of polyolefins is that they are easy to manufacture and process, as well as heat-sealable and cost-effective. PE and PP have respective melting points on the order of 135 and 165 °C. These types of separators typically have high porosities (>40 %) and low ionic resistivity (1.5–2.5 $\Omega \cdot cm^{-2}$), which enable good ionic transport through these thin membranes. The PE and PP separators have reasonably high bulk puncture strengths (>300 $g \cdot mil^{-1}$) [8].

Text Box 3: Note on the use non-SI units

The "g mil^{-1}" unit is one of the examples of a non-SI unit used in manufacturing industry.

(1) In this case, "g" actually means "gram-force"

(2) One mil is equal to one thousandth of an inch, 25.40 μm.
The non-SI units are being used in many sectors of industry. Their usage is mostly custom-based and not always consistent (i.e., "mixed" use of SI/non-SI units).

As the demand for higher energy densities increases, it is desirable to minimize the total quantity of all inert materials in a battery. With a separator thickness of only 20–25 μm, any small intrusion of metallic dust particles can have devastating consequences. High rate cells are looking toward even thinner separators within the range of 9–16 μm to minimize the separator-incurred resistance. Even "gel spots" (binder particles' agglomerates) or bumps in the electrode coatings themselves can puncture through these ultra-thin separators and cause the electrode surfaces to contact one another. Regardless of the cause of the electrical contact, the result is internal cell shorting. If the short-circuit is of a small enough magnitude, it will manifest merely as an elevated self-discharge rate, with little heat generation and low risk. However, if the size of the short is large enough and substantial current is flowing between the electrode plates, the temperature can quickly rise beyond a point where the system is stable and thermal runaway may commence. If the cell's internal temperature approaches or exceeds the PE melting point, even without the commencement of thermal runaway, significant morphological changes occur in separator, resulting from the melting of PE and the swelling of PP [1].

Shutdown separators. Development of the shutdown separators is yet another method to address the cell safety issues. Shutdown separators consist of multiple layers, where at least two of the layers have different phase transition temperatures. As the cell temperature increases, one of the layers begins to melt and flows into the pore structure of the separator. This blocks the ionic path and, ultimately, stops the electrical current flow and thus prevents further heat buildup. The influence of shutdown separators has been debated in the industry. An example of the typical results can be seen in Fig. 5.9. The abscissa is the power rate at which the cell is being discharged during a potential event and the ordinate is the maximum energy that can be drained from the same cell at those power levels. The region

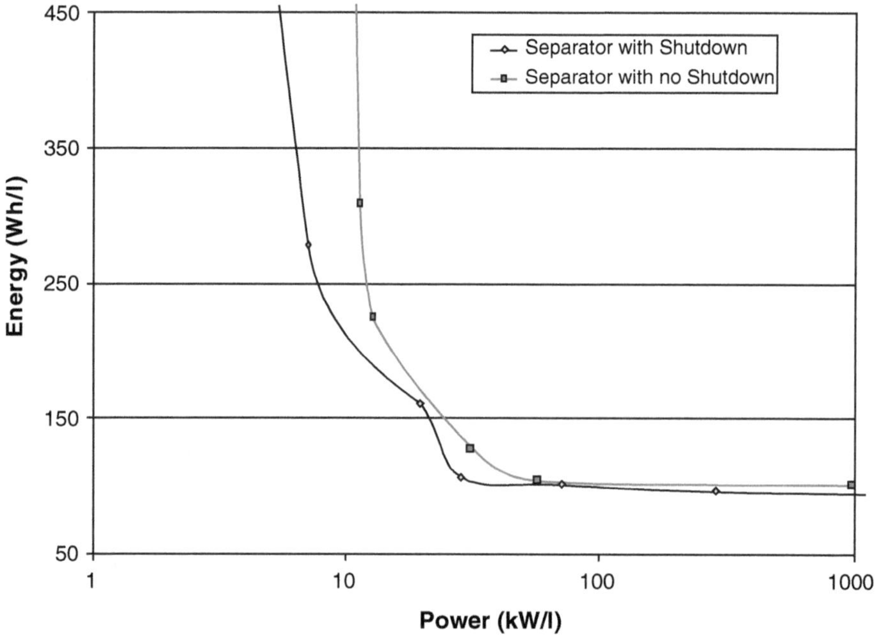

Fig. 5.9 Influence of the shutdown separator on thermal response of a cell

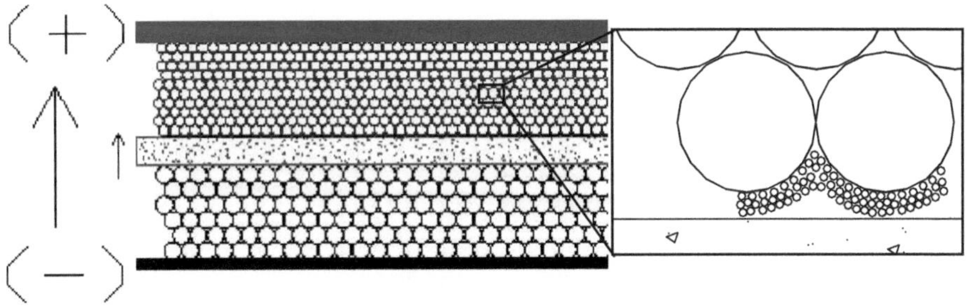

Fig. 5.10 Sketches of the ceramic-coated electrode

above and to the right of each curve represents the area where the cell is likely to enter a thermal runaway situation. In this particular cell design geometry, the use of the shutdown separators affords an additional safety margin of ~10 °C.

One potential area of concern for the safety enhancements via shutdown separators, specifically with large format cells, is the temperature distribution within cells during overcharge or the short-circuit scenarios. The areas of higher temperature can lead to localized separator degradation, shrinkage, or incomplete shutdown. Incomplete shutdown is a scenario where only the pores in the areas of high temperature get filled-in with melted material and this increases the current density in the areas where the pores are still open. These areas now have increased heat generation, further compounding localized heating and degradation and can potentially advance the failure mode of the cell. Areas of degradation and increased shrinkage can leave the electrodes exposed, resulting in regions of internal shorting.

Ceramic coatings on electrodes. Though technically not a separator, ceramic coatings on the surfaces of electrodes help in preventing electrical conduction within the cell, between the cathode and anode when separators begin to fail. A schematic example of a coating on a surface of a cathode is shown in Fig. 5.10.

Although electrochemically inert, separators and ceramic coatings play an integral role in determining the overall safety of lithium-ion cells. The effectiveness of various separator concepts depends on both the size and geometry of the cell and configuration of the battery. Extensive modeling and multiple tests need to be done to evaluate the usefulness of separator technology for a given application (Table 5.2).

5.2 Cell-Level Protection Devices

All the effort focused on the development of the safest possible materials for lithium-ion batteries is somewhat undermined by the tendency to contain ever more energy into ever smaller packages dictated by market needs. Therefore, the safety of lithium-ion cells cannot fully be addressed at the active materials level.

Table 5.2 Selected examples of ceramic coatings on electrodes; notes on various deposition methods used and the effectiveness of coatings [37]

Coating material /substrate	Deposition method(s)	Notes
Li_2O $2B_2O_3$ glass/$LiNi_{0.8}Co_{0.2}O_2$	Solution-based (LiOH and H_3BO_3) followed by heating at 500 °C for 10 h	Coating had high Li^+ conductivity and stability Prevented direct contact between cathode particles and liquid electrolyte [38]
SiO_x/$LiNi_{0.8}Co_{0.2}O_2$	Chemical vapor deposition (CVD)	The coating method was reported as very reproducible but complex in terms of multiple parameters to control Significant (50 %) decrease in the total evolved heat during the exothermic reaction Increased capacity fade was reported on extended cycling [39]
$AlPO_4$/$LiNi_{0.8}Co_{0.1}Mn_{0.1}O_2$	Dispersion in $AlPO_4$-containing slurry followed by drying and the heat treatment	Significantly decreased likelihood of thermal runaway Decreased heat generation in abusive conditions Improved cycle life of the coated cathode [40]
TiO_2/$LiNi_{0.8}Co_{0.2}O_2$	Hydroxylation of tetra-n-butyl titanate (TBT) in air or precipitation from an alcohol solution followed by a heat treatment	Improved cycle life Decreased capacity fade on extended cycling Suppressed decomposition of electrolyte in abusive conditions [41]

Fig. 5.11 Cross-sectional diagram of 18,650 cell showing CID and PTC

A complementary approach at the cell design level needs to be taken to enhance safety characteristics. In principle, the cell failures may not be absolutely eliminated but their probability should be significantly reduced and the severity of such events should be mitigated and their impact limited.

In addition to the selection of the cell materials (both active and inactive) the cell design greatly impacts the battery safety, providing the opportunity for the installation of the cell-level protection devices. One such device is a disk with a positive temperature coefficient of expansion often referred to as a PTC. PTCs are used extensively in cylindrical cells like the commonplace 18,650 (this common cylindrical cell designation references the cell dimensions, where the cell base diameter is 18 mm and the cell height equals 65.0 mm).

The PTC is a thin annulus consisting of the polyethylene-based matrix, with conductive materials dispersed throughout [42]. The PTCs are placed in the current path within the cell, typically being embedded in the header as shown in Fig. 5.11. When experiencing an overcurrent situation as would be the case in the event of an external short, the polymer heats up and expands. The polymer expansion causes the increase of electrical resistivity of the PTC and the resulting decrease of the current flow through the cell. If the external short is removed, the PTC has the chance to cool and contract to the original state, causing the cell's resistance to diminish. While PTCs certainly perform their intended function in certain applications, they have several weaknesses that prevent them from being used in every cell. PTCs are not suitable for use in high rate cells, where maintaining the lowest possible total resistance is the main goal. Likewise, they are not appropriate for use in high voltage applications. When a PTC trips in a high-voltage battery configuration, voltage drop can be experienced that exceeds the safe battery working voltage. Tests performed at COM DEV International and at NASA–Johnson Space Center have revealed that sparks and flames accompany these PTC failures [42]. In an over-voltage condition, the polyethylene may char and form a low resistance short, a scenario that the PTC was designed to prevent in the first place. This could then exacerbate the event by overheating and melting the header gasket, and, eventually, shorting the can (negative terminal) to the cover (positive terminal). The short-circuit path thus created through the cell can lead to thermal runaway.

In addition to PTCs, there is another group of the cell-level protection devices typically installed in small cylindrical cells, such as 18,650s. These cell safety-enhancing accessories, which take advantage of the excess pressure generated in the cell during the early stages of thermal runaway, are known as the current interruption devices (CID). CIDs are mechanical switches incorporated into the cap assembly of the cell much like PTCs (Fig. 5.11). The positive electrode tab is connected through the CID to the positive terminal of the cell. In case of the excessive pressure increase during high voltage charging, the CID breaks the connection between the terminal and the positive electrode, opening the electrical path and preventing any additional charge from entering the cell preventing overcharge and violent cell failures. While CIDs can help mitigate failure of the cell due to overcharging and external shorting, they offer little help for other failures (Fig. 5.1). They are mostly found in high energy cells due to the difficulty in developing a CID that can break high currents.

The excessive pressure buildup in cells, whether it is from aging, overtemperature or overcharge tends to expedite the chemical reactions within the cell. This can promote the onset of thermal runaway. The excess pressure can also lead to mechanical failures distorting or even rupturing the cell casing in a catastrophic fire or explosion. To mitigate this effect, design engineers also incorporate rupture disks or cell vents into the cell design to control the release of the gases before the pressure builds up to dangerous levels. These are typically designed as scored portions of the cell casing, producing a weakened section of the can to control the pressure and the direction of effluent expulsion. The cell-level protection devices exhibit a reasonable level of reliability, resulting (when combined with other safety-enhancing strategies) in a relatively low statistical failure rate of lithium-ion cells in the field (estimated 1 in 10 million cells). Still, due to the highly energetic nature of lithium-ion cells and batteries, their failures are rather memorable, if sparse.

5.3 Safety at the Battery Level

With energy densities of current generation high-end lithium-ion cells reaching and exceeding 200 $Wh \cdot kg^{-1}$, the cells contain enough electrochemical energy to adiabatically heat themselves and increase temperature by >660 °C. This temperature is well above the thermal stability of the active components and results in significant additional chemical energy release from the decomposition of the cathode, anode, and burning of the electrolyte. The vigor with which the chemical energy is released adds to the propensity for propagation of the single cell event onto further, neighboring cells. There will always be certain failure modes that would result in the rapid release of the stored energy; this is nearly unavoidable. The key to a safer battery is:

(1) Reducing the probability of the single cell failure propagation leading to the complete battery failure.
(2) Diminishing the severity of any further cell failures that may occur.

The results of various safety tests, discussed more closely in the latter section of this chapter are used to realistically assess the safety of the entire battery treated as a system having different thermal properties and behavior characteristics under abusive conditions than a single cell. Failure to properly prevent cell-to-cell event propagation jeopardizes entire missions and can put lives in danger.

5.3.1 Thermal Management

Data available from the cell-level and pack-level safety tests in the form of thermal runaway onset temperature, heat generation rates, cell face temperatures, and gas generation rates/composition needs to be analyzed by the engineers so they can develop a suitable thermal network. Other limitations that do not necessarily directly affect the safety of the battery system need to be considered as well prior to designing the thermal management system. These include allowable temperature gradient across the battery, allotted space within the application to house the battery, and available power sources. Temperature variations within a battery could lead to different current drains (i.e., discharge current rates) for each of the cells. Ultimately, this may lead to state of charge imbalance, reduce pack performance and, as will be discussed later, potentially lead to overcharge conditions.

Depending on the type of the thermal system and their usage and thermal output profiles, there are several different cooling methods that can be utilized, as illustrated in Fig. 5.12. The active and the passive cooling methods are two main categories of the cooling techniques available for use in thermal management of batteries. These cooling methods are suited to different conditions and analysis is required to determine which option is appropriate. There are instances where both methods are recommended. Passive cooling designs rely on using solely the three fundamental heat transfer methods (conduction, convection, and radiation) and do not employ fans, pumps, heat exchangers, or compressors, unlike the actively cooled batteries.

Currently used nickel–metal hydride (Ni–MH) battery packs in Honda Insight and Toyota Prius provide a good example of the active air cooling. Possible application of similar systems to lithium-ion batteries was discussed by Bandhauer et al. [44], where many challenges ahead of the successful thermal management in lithium-ion batteries were highlighted. Li-ion batteries are more energy dense than Ni–MH packs and therein lies the problem with a direct transfer of cooling methods between these two technologies [43]. Liquid-based active cooling [44]

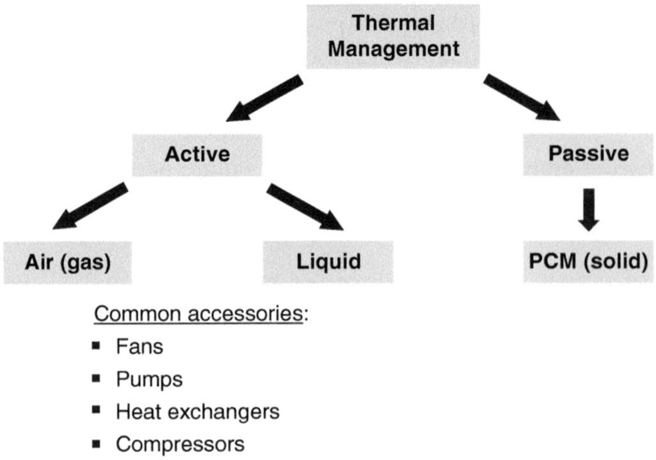

Fig. 5.12 The most common thermal management methods. Note: passive air or liquid cooling is also possible, although uncommon

or passive cooling based on the use of the solid phase change material (PCM) [43] were suggested as the alternative, higher performing thermal management methods.

Compared to air, liquids can be used more efficiently for active cooling. The large heat capacity and high density of liquids allow transmitting heat with reduced temperature difference and much less volumetric flow than in case of air used as coolant. However, problems may arise if coolant is unstable in certain conditions that may be generated in the presence of the battery. For example, water-based coolant may undergo electrolysis above 1 V, producing hydrogen gas if placed in contact with both electrodes of the battery [44].

The PCMs utilize latent heat of melting/solidification processes to absorb and pull away heat from failing cells thus minimizing the risk of reaching thermal runaway conditions [43, 44]. The development of PCMs is relatively new and the testing relies much on simulations of the heat flow under various conditions. An example of the PCM is paraffin wax used with or without addition of expanded metal foam component. From the simulation and initial testing results, PCMs are attractive alternative for air or fluid cooling but several disadvantages were also identified:

(1) PCMs increase the volume and weight of the battery pack; the most disadvantageous volume/weight gain is seen when PCM is combined with an active cooling method (air or liquid)
(2) The complete meltdown of PCM in the dual cooling system (e.g., PCM + active fluid cooling) would form a thermal barrier between the coolant, thus decreasing the efficiency of the active cooling component
(3) In colder than ambient conditions (e.g., encountered by electric vehicles during winter driving), PCMs would interfere with the attempts at warm-up of the battery (e.g., by external heater) and potentially induce the temperature gradient amongst the cells.

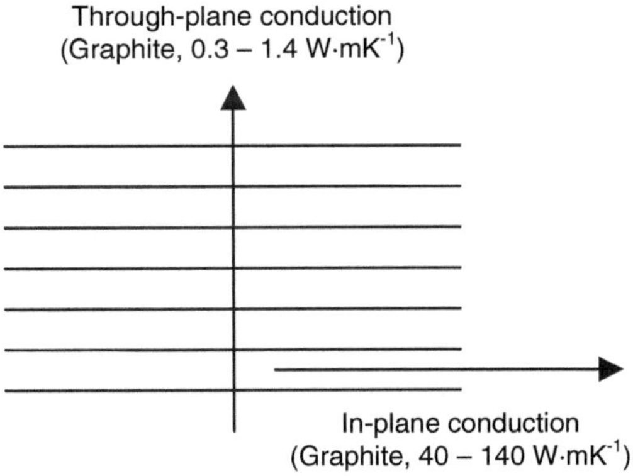

Fig. 5.13 Assessment of the thermal properties of the cell materials and/or cells

In order to accurately evaluate the effectiveness of the established and emerging thermal management techniques, it is essential that the thermal and mechanical properties of all of the battery components must be fully known. The thermal properties of the cells, typically the largest contributing component in the battery, both by mass and volume, are most often the least understood of all components.

Figure 5.13 depicts an apparatus to experimentally measure the thermal conductivity of prismatic lithium-ion cells in three dimensions. The aluminum and copper substrates that the electrodes are coated on (i.e., the current collecting foils) may have thermal conductivities that are orders of magnitude higher than the coatings themselves. Specifically, the majority of cathode materials have extremely poor thermal conductivities (e.g., $LiCoO_2$ 1.58 W mK^{-1}), whereas aluminum and copper, commonly used for current collectors, are 240 W mK^{-1} and 395 W mK^{-1}, respectively [45]. The carbon-based anode materials are not much better than the cathodes, though the graphitic or silicon-based anodes are much more comparable to the substrates themselves in terms of their heat conduction properties. This accounts for a large difference in the thermal conductivity measured in the plane of the electrodes from that measured through the plane of the electrodes. Further, it is essential to understand the specific heat capacity of the cell as well, which can be attained from the traditional calorimetry experiments.

An understanding of the specific heat capacity of the cells en masse is critical when it comes to mitigation of the cell-to-cell event propagation. The mitigation is usually done by insuring that rate of heat removal is greater than rate of heat generation within the cells. To this effect, it is important to develop thermal network connections to increase the immediately accessible thermal mass to limit temperature of adjacent cells and to spread it out across as many cells as possible. Thus, the neighboring cells are prevented from exceeding the thermal runaway threshold. The runaway threshold temperature is difficult to define as a single number; some researchers have set this limit at 60 °C [4] while others—above 70 °C [45]. This is not a surprising discrepancy, considering the multitude of cell chemistry options and cell designs that are being tested for their thermal response to the abusive treatment. The test conditions, such as the cathode's state of charge or the cell's cycling history also have a great impact on the thermal characteristics of the battery.

When defining the optimum internal operating temperature range for lithium-ion cells, several (sometimes mutually contradictory) factors must be considered. For example the best balance between long lifetime and cell performance is achieved when the mean temperature lies between 25 and 30 °C, but the cell performance generally improves with increasing temperature, as Li-ion diffusion rate is accelerated. However, a significant reduction of cell life is observed at elevated temperatures. As much as 50 % decrease in cell life expectancy was reported for a temperature increase of 10 °C [45]. Therefore, the successful thermal management system must ensure a long battery lifetime, enable good performance characteristics but also take the battery's safety into consideration.

A complex, multivariable problem, such as designing the optimum thermal management system for a lithium-ion battery requires sophisticated calculating approaches. Finite Element Analysis (FEA) software has been commonly used to assess whether the proposed battery design would perform as expected and meet all requirements, both thermally and mechanically. FEA is especially useful in solving problems where complex geometries are concerned [45]. The numerical methods that are the foundation of FEA are based on solving multiple, relatively simple equations that describe small subdomains (finite elements), to approximate the more complex equation for a larger domain. In graphical representation, a grid (mesh) mesh is created that covers the complex object (e.g., battery and the cooling system) for which the calculations are done. Thermal and/or structural properties of the system are defined for each node point of the mesh. Vectors connecting the nodes are used to describe the directional change in properties of a given subdomain. The most popular use of FEA to date has been to predict the mechanical behavior of complex systems (e.g., car body design) and to predict their fatigue or failure modes. When the numerical values of thermal properties (e.g., specific heat coefficients, thermal conductivity coefficients, etc.) are loaded into the model, the thermal analysis can be performed by FEA software.

The example of an FEA being used for battery safety improvements is the determination of the lithium-ion cell aspect ratio that is best for transporting the heat away from a failing cell but not toward the neighboring cells. Based upon the FEA results, thinner cells were identified as better candidates to move heat to edges faster (Fig. 5.14). Cells of lower total energy (used in appropriately larger numbers) were found beneficial for minimizing total heat and gas evolved in the simulated event.

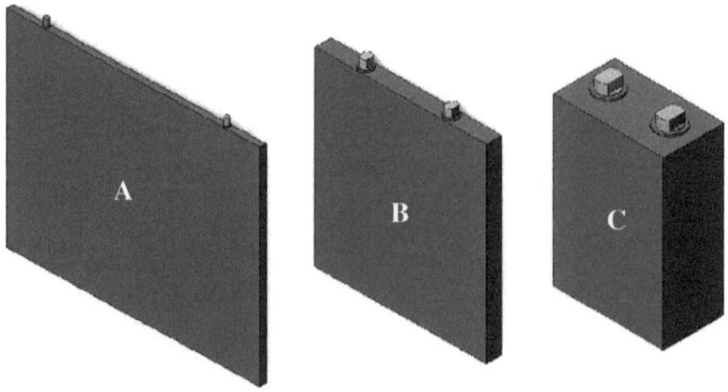

Fig. 5.14 Three cell design options, varying in aspect ratios and the resulting thermal properties

5.3.2 Battery Management System (BMS)

Lithium-ion cells can be damaged and may catastrophically fail resulting in venting flames or explosion when used outside of manufacturer specified operating temperature ranges. Cells may also be damaged if they are allowed to be charged above or discharged below specified safe operating voltage. For this reason lithium-ion batteries require a BMS. The BMS is used to prevent cells from cycling outside of the safe operating area and to shut down the battery pack in the event of cell or BMS failure. The safe operating conditions can be exceeded by the influence of such factors as small manufacturing variances in capacity, cell's resistance, and/or self-discharge as well as environmental influences or misuse.

The BMS role in battery is based on the following tasks:

(1) Monitoring temperature
(2) Monitoring voltage
(3) Controlling charging process and charging contactors
(4) Equalization and balancing of cells
(5) Rerouting excess energy
(6) Short-circuit protection
(7) Control of active cooling equipment

In battery packs with more than one series of connected cells, it is possible for the voltage of individual cells to become mismatched (Fig. 5.15). This could be due to a variety of reasons, from the aforementioned manufacturing variances to physical location of a cell in the battery. Areas inside the battery housing may have higher local temperatures and the aging characteristics of cells located in these areas will differ from cells mounted in cooler positions within the string. This mismatch could cause premature cells' capacity degradation and increased resistance of the cells. The

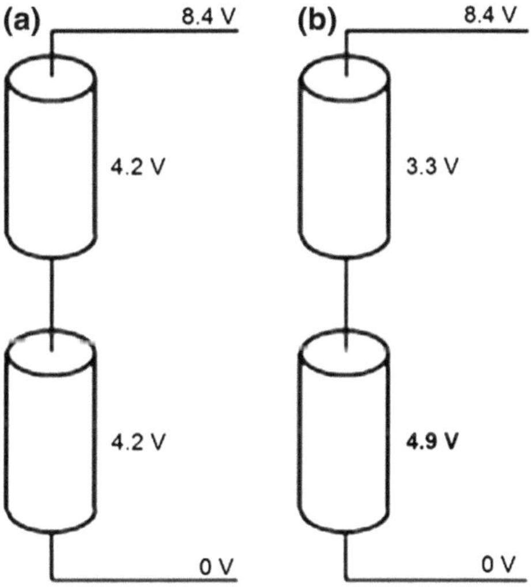

Fig. 5.15 Two strings of cells connected in series **a** balanced and **b** unbalanced

combination of the above-mentioned factors can lead to voltage imbalance between cells. When connected to a power supply for charging, the supply "sees" only the total battery voltage (which is the summation of all individual series-connected cells). Therefore, the BMS is necessary to monitor each individual cell voltage and to keep every cell within the safe operating voltage window. If the safe operating voltage limits of any cell are exceeded, the BMS must terminate the charge, either by directly turning off the charger or by opening a circuit-breaking switch, thus breaching any electrical paths between the charger and the battery.

For the optimal operation of the BMS (and the battery), all cells should exhibit close operating voltage values, i.e., the cell pack must be balanced. Charging of an unbalanced pack could cause one cell to trigger the termination of the battery charging before all the cells reach their full charge state. During discharge, these undercharged cells would reach their under-voltage cutoff limit and, in turn, terminate discharge. Obviously, this sequence of events is undesired as the capacity of the battery pack would thus be greatly diminished.

In the hypothetical situation when one cell reaches the upper voltage limit prior to the rest of the pack, the BMS cannot simply allow the charger to increase the voltage of the pack for the rest of the cells to "catch up". In such event, one cell would exceed the safe voltage limit and could potentially fail catastrophically.

The above scenarios illustrate the reason why the BMSs must "balance" or "equalize" the cells during the charging process. There are several ways that this balancing can be accomplished, with each approach having drawbacks and benefits. The most widespread commercial balancing method is resistive bypass. When the bypass method is employed, each cell is connected to a resistor through a switch controlled by the BMS. As the cells with higher voltage approach their over-voltage threshold, the switch is closed and these cells are discharged through their respective resistors. Effectively, these cells will receive lower charge current than cells not discharged through the resistor. This passive balancing technique suffers from overall low energy efficiency and the related heat generation at the sources of the resistance.

Higher energy efficiency, active balancing methods have been developed for batteries with a relatively small number of cells connected in series. "Charge shuttling" methods operate on the principle of transferring charge from the higher voltage cells to a charge storage medium (capacitor or inductor). The BMS would then route this charge to an adjacent cell with a lower voltage. This method has much higher overall energy efficiency, as excess energy is rerouted and not wasted in the form of heat. However, in batteries with long cell strings it is possible for the high and low voltage cells to be on opposite sides of the battery. This arrangement is ineffective as the switching circuitry grows in complexity and there is difficulty associated with cascading the excess charge through all cells connected in series, until reaching the destination cell. Other active balancing methods require the use of transformers [46]. One major benefit of active balancing is that the pack is no longer limited in capacity by the weakest cell. The active balancing methods are not limited to the exclusive operation during charging; they can provide additional charge to the "weakest" cells during discharge, thus extending the discharge capacity of the entire pack (Fig. 5.16).

The accurate operation of BMS relies on a network of sensors spread throughout the battery. All series-connected cells must have their voltage individually

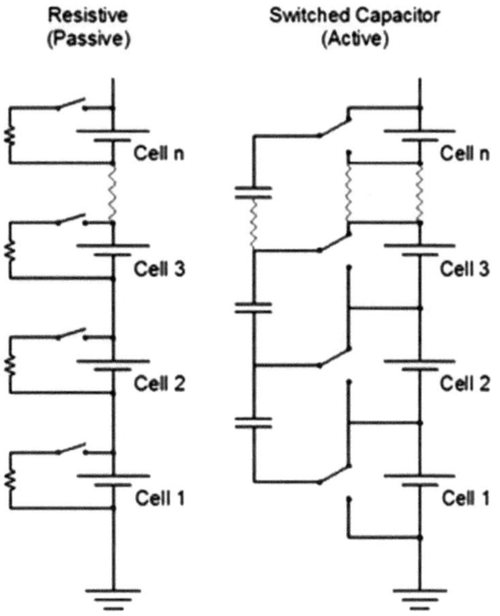

Fig. 5.16 Examples of the passive (*left*) and active (*right*) balancing arrangements in lithium-ion battery

monitored to ensure their usage within the safe voltage window. Only one voltage sensor is required to monitor multiple cells connected in parallel. This is because the parallel-connected cells are, by definition, at the same voltage.

The principal function of the BMS is to protect the battery pack from over-voltage (terminate charge) and under-voltage (terminate discharge) and to prevent damage to the battery, while reducing the risk of catastrophic failures. Since the relationship between voltage and capacity can be easily determined for most lithium-ion batteries,[2] certain BMSs may use voltage sensors to provide for a "fuel gauge" estimate of the capacity remaining in the battery.

In addition to voltage probes, temperature sensors must also be incorporated into the cell pack at specific locations. The need for localized temperature sensing is based on the fact that there is a significant (and sometimes, nonuniform) generation of heat during the charging of battery and, as the temperatures increases, so does the likelihood of entering the thermal runaway conditions. As discussed in a previous section, FEA should be performed at the battery design stage to determine the number and locations of all temperature sensors required to detect if any cell is nearing thermal runaway temperatures.

The BMS would also be in command of any active cooling techniques employed in the battery to allow for operation of the battery even when ambient conditions

[2] The relationship between battery voltage and the remaining capacity is not easily established in lithium-ion batteries that are based on the lithiated metal phosphate cathodes (e.g. LiFe-PO4, lithium iron phosphate). This is due to the fact that the voltage profile for phosphate cathode is flat for almost entire range of the battery's state of charge. It is much easier to correlate the voltage vs. capacity relationship when the voltage profile is a sloping curve (e.g. for LiNi0.8Co0.2O2 cathodes).

fall outside of the recommended operating windows. On a battery level, thermostats are often used to prevent charging at temperatures that are beyond certain threshold values. Additionally, the BMS contains sensors that monitor the current that is coming into or out of the battery. These sensors allow the BMS to communicate and control the chargers to enable the optimal battery charging routine. When the battery is at a low state of charge, the charger should provide maximum charging current allowable by the manufacturer. As the battery reaches its maximum cutoff voltage, the current should be tapered down to maintain the battery at this constant voltage. This procedure is referred to as CC/CV charging profile. The CC/CV profile allows for maximum capacity to be delivered into the cell and for optimum efficiency of the balancing circuitry during the CV portion of the charge.

The BMS will also take action if high currents are detected that may damage the battery. The BMS may command a contactor or relay to open and remove these short-circuits to protect the battery from overheating. Current sensors with high accuracy can also be used for coulomb counting as a more accurate method of "fuel gauging". The coulomb counting method is commonly used to estimate the remaining capacity in $LiFePO_4$-based batteries.

5.4 Safety Testing

Regardless of their design differences, all cell and/or battery types must undergo extensive safety testing prior to their introduction to the market. Safety testing is necessary to confirm the predicted behavior of the device in case of the experienced failure. A list of safety tests routinely performed on lithium-ion batteries is available from Spotnitz and Franklin [2, 47]. Few selected safety tests are described below.

Overcharge test. During the overcharge test, electrical current is forced through the cell up to the certain predetermined voltage limit. The electrochemical reactions within the cell (e.g., Eq. 5.1) and the flow of current through the cell generate heat that is measured and recorded for comparison purposes.

Nail penetration. In this test, a nail is driven through the battery (or a cell) at a given, constant rate (e.g., 8 cm s^{-1}). Electrical current flows through the cell (or battery), and through the nail once a direct short is caused by the nail connecting the adjacent electrode pairs that normally would not be in contact. Total heat generated by combined flow of current is tracked and documented.

Crush test. A bar is used to press down on the battery until the short-circuit is initiated. Generated heat originates from the battery due to induced discharge, from the electrode pairs that are short-circuited.

5.4.1 Safety Testing Examples

Examples of the selected abuse testing are given on the single cell level as well as for the pack level.

5.4.1.1 Test Example 1: Gross Overcharge and Event Propagation

During this gross overcharge/event propagation test presented below, one cell was actively overcharged in a five-cell pack, with temperature and voltage monitoring probes attached to multiple points of the pack. Two five-cell packs were tested: one without heat mitigation and the second pack with the heat-mitigating insulation material. The heat-mitigating strategies were developed based on the results of thermal modeling, discussed in the previous section of this chapter.

Several cells were mounted between two plates of similar heat capacity and one cell (middle cell—Cell #3, as seen in Fig. 5.17) was grossly overcharged in order to initiate the cell failure. The remaining cells were all at 100 % SOC. This test determined the heat flow to the neighboring cells and likelihood of the event propagation. In both cell packs, Cell #3 was actively overcharged and vented at about 700 % SOC in the unmitigated pack and 500 % SOC in the mitigated pack, respectively. The observed individual differences between the grossly overcharged cells and their SOC can be very large and the exact moment of failure is rather difficult to predict. Under the gross overcharge conditions, 200 % SOC difference can be attributed to such random factors as the small particle agglomerate on one of the electrodes that could affect the local current density. Because of the wide distribution of the gross overcharge testing results, more cells need to be tested to properly quantify the average SOC at which the overcharge triggers the thermal runaway. However, the qualitative conclusions can be drawn on the total thermal effect on the neighboring cells. In this test, the impact of the failing cell on its neighbors differed greatly, depending on whether the heat-mitigating material was used in the cell pack.

In the unmitigated pack, after Cell #3 has been forced into the gross overcharge, there was no catastrophic propagation onto the other cells and no fire; however, all cells have vented and lost voltage as a result of this test. The thermal effect was reasonably close in all cells and the measured cell temperature was

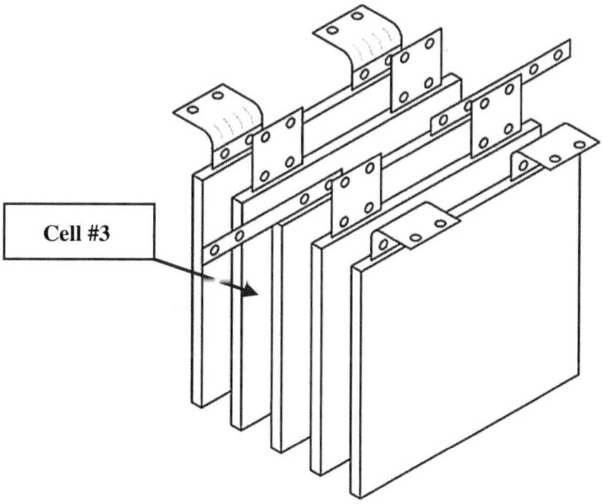

Fig. 5.17 Five-cell pack gross overcharge test setup

about ~285 °C. In the heat-mitigated cell pack, Cell #3 was driven to gross overcharge and then vented, but the remaining cells retained their voltage of 4.1 V.

The thermal measurements have indicated that Cell #3 was at temperature of ~173 °C at the time of venting and the remaining cells were all at the temperature of 90 °C or below, which is the reason why they have remained operational (Cell #3, obviously, failed). In addition to the neighboring cells having the maximum temperature of 90 °C or below at the time of Cell #3 venting, all these cells were brought down to temperature ~50 °C within 0.5 h from the gross overcharge event. One may speculate that if the heat mitigation system was to be further optimized, even further lowering of the neighboring cells temperature at the time of event may be possible, as well as the faster heat dissipation away from the cell pack.

5.4.1.2 Test Example 2: Nail Penetration

In the nail penetration test, a lithium-ion cell was mounted between two plates of similar heat capacity and nail was driven into the cell while the cell was fully charged (100 % SOC) in order to initiate an internal short-circuit. The test typically determines the heat flow immediately after the nail penetration. In this case, the temperature of the cell is monitored until the ambient value is reached (Fig. 5.18).

Within 300 s from the nail penetration, the cell's external case temperature reached 200–260 °C temperature (differences are caused by probes' mounting spots). It took about an hour for the cell to cool down to 50 °C and about another hour to reach ambient temperature.

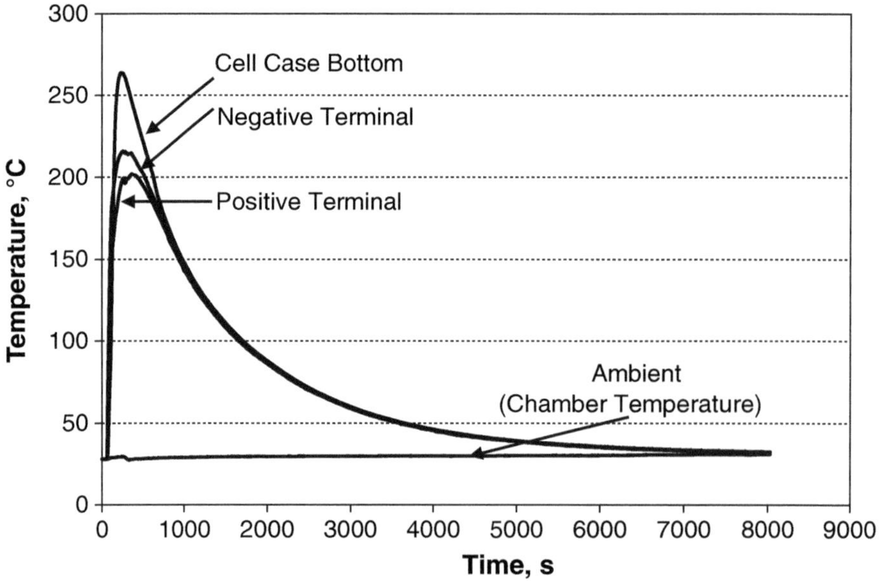

Fig. 5.18 Single cell nail penetration: temperature plot versus time

5.5 Conclusions

Safety-related cell and battery design challenges are most likely to be the leading topic for research and development in the next 5–10 years. The field of the cell chemistry development remains very prolific and new materials for lithium-ion battery applications are reported almost every day. Sometimes a subtle change in material composition, such as a few percent of doped ions (e.g., aluminum or magnesium introduced into $LiNi_{0.8}Co_{0.2}O_2$ cathode) can have a serious impact on the safety properties of the cell. A multitude of testing methods is available for assistance with testing of materials, components, cells, and cell packs. More standardization of the testing methods and conditions is desired for more linear progress in battery safety. Another much desired development would be a comprehensive database of materials, their properties, and their expected behavior in cells of various designs (true prismatic, cylindrical, polymer cells, etc.).

The safety of the system is analytically determined from the joint contributions of all components. Modeling of thermal and mechanical properties of batteries remains a useful method to approximate their systemic behavior. However, the experimental verification and the (expensive) destructive testing still provide the most valuable data on the actual event progress and propagation. Improvements in the accuracy of thermal models as well as the ever-increasing wealth of data from the past and ongoing testing are expected to improve the reliability of the cell performance predictions.

5.6 Further Reading

1. Gnanaraj JS et al. (2003) A detailed investigation of the thermal reactions of $LiPF_6$ solution in organic carbonates using ARC and DSC. J Electrochem Soc 150:A1533–A191. doi:10.1149/1.1617301
2. Doughty D, Roth P (2012) A general discussion of Li ion battery safety. Electrochem Soc Interface 21:37–44
3. http://www.technologyreview.com/news/416973/cheaper-stronger-lithium-ion-batteries-for-electric-vehicles/. Accessed May 21
4. http://cen.acs.org/articles/91/i6/Assessing-Safety-Lithium-Ion-Batteries.html. Accessed May 21
5. http://www.eurekalert.org/pub_releases/2013-04/acs-lbt032113.php. Accessed May 21
6. Orendorff CJ (2012) The role of separators in Lithium-ion safety. Electrochem Soc Interface 21:61–65
7. Maleki H, Deng GP, Anani A, Howard J (1999) Thermal stability studies of Li-ion cells and components J Electrochem Soc 146:3224–3229. doi: 10.1149/1.1392458

References

1. Abraham DP et al (2006) Diagnostic examination of thermally abused high-power lithium-ion cells. J Power Sources 161:648–657. doi: 10.1016/j.jpowsour.2006.04.088
2. Lisbona D, Snee T (2011) A review of hazards associated with primary lithium and lithium-ion batteries. Process Saf Environ 89:434–442. doi: 10.1016/j.psep.2011.06.022
3. Balakrishnan PG, Ramesh R, Prem KT (2006) Safety mechanisms in lithium-ion batteries. J Power Sources 155:401–414. doi: 10.1016/j.jpowsour.2005.12.002
4. Ohsaki T (2005) Overcharge reaction of lithium-ion batteries. J Power Sources 146:97–100. doi: 10.1016/j.jpowsour.2005.03.105
5. Sone Y et al (2007) Charge and discharge performance of over-discharged lithium-ion secondary battery - Lessons learned from the operation of the interplanetary spacecraft HAYABUSA. Electrochemistry 75:950–957
6. Hall JC et al (2006) Decay processes and life predictions for lithium ion satellite cells. 4th International Energy Conversion Engineering Conference and Exhibit (IECEC) 26-29 June 2006, San Diego, California
7. Sabbah R (1999) Thermochim Acta 331:93–204. doi: 10.1016/S0040-6031(99)00009-X
8. Roth EP, Crafts CC, Doughty DH, McBreen J (2004) Advanced technology development program for lithium-ion batteries: thermal abuse performance of 18,650 Li-ion cells. Sandia National Laboratories SAND2004-0584 report
9. MacNeil DD, Lu Z, Chen Z, Dahn JR (2002) A comparison of the cathode/electrolyte reaction at elevated temperatures for various Li-ion battery cathodes. J Power Sources 108:8–14. doi: 10.1016/S0378-7753(01)01013-8
10. Whittingham MS (2004) Lithium batteries and cathode materials. Chem Rev 104:4271–4301. doi: 10.1021/cr020731c
11. Wen CY et al (2012) Thermal runaway features of 18650 lithium-ion batteries for LiFePO4 cathode material by DSC and VSP2. J Therm Anal Calorim 109:1297–1302.doi: 10.1007/s10973-012-2573-2
12. Aurbach D et al (1994) The correlation between the surface-chemistry and the performance of li-carbon intercalation anodes for rechargeable rocking-chair type batteries. Electrochem Soc 141:603-611. doi: 10.1149/1.2054777
13. Aurbach D et al (1996) A comparative study of synthetic graphite and Li electrodes in electrolyte solutions based on ethylene carbonate dimethyl carbonate mixtures. J Electrochem Soc 143:3809–3820. doi: 10.1149/1.1837300
14. Peled E, Golodnitski D, Menachem C, Bar-Tow D (1998) An advanced tool for the selection of electrolyte components for rechargeable lithium batteries. J Electrochem Soc 3482–3486. doi: 10.1149/1.1838831
15. Imhof R, Novak P (1998) In situ investigation of the electrochemical reduction of carbonate electrolyte solutions at graphite electrodes. J Electrochem Soc 145:1081–1087. doi: 10.1149/1.1838420
16. Chagnes A, Swiatowska J (2012) Electrolyte and solid-electrolyte interphase layer in lithium-ion batteries. In: Belharouak I (ed) Lithium ion batteries—new developments. InTech, Rijeka, Croatia. doi: 10.5772/1358
17. Yang H, Bang H, Amine K, Prakash J (2005) Investigations of the exothermic reactions of natural graphite anode for Li-ion batteries during thermal runaway. J Electrochem Soc 152:A73–A79. doi: 10.1149/1.1836126
18. Campion CL et al (2004) Suppression of toxic compounds produced in the decomposition of lithium-ion battery electrolytes. Electrochem Solid-State Lett 7:A194–A197. doi: 10.1149/1.1738551
19. Wang XM et al (2006) High-concentration trimethyl phosphate-based nonflammable electrolytes with improved charge-discharge performance of a graphite anode for lithium-ion cells. J Electrochem Soc 153:A135–A139. doi: 10.1149/1.2136078

20. Wang X, Yasukawa E, Kasuya S (2001) Nonflammable trimethyl phosphate solvent-containing electrolytes for lithium-ion batteries–I. Fundamental properties. J Electrochem Soc 148:A1058–A1065. doi: 10.1149/1.1397773
21. Xu K et al (2002) An attempt to formulate nonflammable lithium ion electrolytes with alkyl phosphates and phosphazenes. J Electrochem Soc 149:A622–A626. doi: 10.1149/1.1467946
22. Xu K et al (2003) Evaluation of fluorinated alkyl phosphates as flame retardants in electrolytes for Li-ion batteries–I. Physical and electrochemical properties. J Electrochem Soc 150:A161–A169. doi: 10.1149/1.1533040
23. Xu K, Zhang SS, Allen JL, Jow TR (2003) Evaluation of fluorinated alkyl phosphates as flame retardants in electrolytes for Li-ion batteries–II. Performance in cell. J Electrochem Soc 150:A170–A175. doi: 10.1149/1.1533041
24. Izquierdo-Gonzales S; Li W; Lucht BL (2004) Hexamethylphosphoramide as a flame retarding additive for lithium-ion battery electrolytes. J Power Sources 135:291–296. doi: 10.1016/j.jpowsour.2004.04.011
25. Xiang HF; Xu HY; Wang ZZ; Chen CH (2007) Dimethyl methylphosphonate (DMMP) as an efficient flame retardant additive for the lithium-ion battery electrolytes. J Power Sources 173:562–564. doi: 10.1016/j.jpowsour.2007.05.001
26. Xiang HF et al (2007) Dimethyl methylphosphonate-based nonflammable electrolyte and high safety lithium-ion batteries. J Power Sources 174:335–341. doi: 10.1016/j.jpowsour.2007.09.025
27. Feng JK, Ai XP, Cao YL, Yang HX (2008) Possible use of non-flammable phosphonate ethers as pure electrolyte solvent for lithium batteries. J Power Sources 177:194–198. doi: 10.1016/j.jpowsour.2007.10.084
28. Zhang SS (2006) A review on electrolyte additives for lithium-ion batteries. J Power Sources 162:1379–1394. doi: 10.1016/j.jpowsour.2006.07.074
29. Broussely M et al (2005) Main aging mechanisms in Li ion batteries. J Power Sources 146:90–96. doi: 10.1016/j.jpowsour.2005.03.172
30. Xu K, Lee U, Zhang S, Wood M, Jow TR (2003) Chemical analysis of graphite/electrolyte interface formed in LiBOB-based electrolytes. Electrochem Solid-State Lett 6:A144–A148. doi: 10.1149/1.1576049
31. Jiang J, Dahn JR (2004) Effects of solvents and salts on the thermal stability of LiC6. Electrochim Acta 49:4599–4604. doi: 10.1016/j.electacta.2004.05.014
32. Li W et al (2005) Additives for stabilizing LiPF6-based electrolytes against thermal decomposition. J Electrochem Soc 152:A1361–A1365. doi: 10.1149/1.1926651
33. Xiao A, Lucht BL (2006) Thermal reactions of mesocarbon microbead (MCMB) particles in LiPF6-based electrolyte. J Power Sources 162:1282–1288. doi: 10.1016/j.jpowsour.2006.07.054
34. Li W, Lucht BL (2006) Lithium-ion batteries: Thermal reactions of electrolyte with the surface of metal oxide cathode particles. J Electrochem Soc 153:A1617–A1625. doi: 10.1149/1.2210588
35. Abe K, Ushigoe Y, Yoshitake H, Yoshio M (2006) Functional electrolytes: Novel type additives for cathode materials, providing high cycleability performance. J Power Sources 153:328–335. doi: 10.1016/j.jpowsour.2005.05.067
36. Dias FB, Plomp L, Veldhuis JBJ (2000) Trends in polymer electrolytes for secondary lithium batteries. J Power Sources 88:169–191. doi: 10.1016/S0378-7753(99)00529-7
37. Li C et al (2006) Cathode materials modified by surface coating for lithium ion batteries. Electrochim Acta 51:3872–3883. doi: 10.1016/j.electacta.2005.11.015
38. Ying JR, Wan CR, Jiang CY (2001) Surface treatment of LiNi0.8Co0.2O2 cathode material for lithium secondary batteries. J Power Sources 102:162–166. doi: 10.1016/S0378-7753(01)00795-9
39. Omanda H et al (2004) Improvement of the thermal stability of $LiNi_{0.8}Co_{0.2}O_2$ cathode by a SiO_x protective coating. J Electrochem Soc 151:A922–A929. doi: 10.1149/1.1710892
40. Cho J et al (2004) Synthesis, thermal, and electrochemical properties of AlPO4-coated LiNi0.8Co0.1Mn0.1O2 cathode materials for a Li-ion cell. J Electrochem Soc 151:A1899–A1904. doi: 10.1149/1.1802411

41. Zhang ZR et al (2004) Electrochemical performance and spectroscopic characterization of TiO2-coated LiNi0.8CO0.2O2 cathode materials. J Power Sources 129:101–106. doi: 10.1016/j.jpowsour.2003.11.015
42. Darcy E, Davies F, Jeevarajan J, Cowles P (2003) Lithium-ion Cell PTC limitations and solutions for high voltage battery applications. The Electrochemical Society 203rd Meeting, April 27–May 2, 2003, Paris, France. Abstract #189
43. Kizilel R, Sabbaha R, Selmana JR, Al-Hallaj S (2009) An alternative cooling system to enhance the safety of Li-ion battery packs. J Power Sources 194:1105–1112. doi: 10.1016/j.jpowsour.2009.06.074
44. Bandhauer TM, Garimella S, Fuller TF (2011) A critical review of thermal issues in lithium-ion batteries. J Electrochem Soc 158:R1–R25. doi: 10.1149/1.3515880
45. Kruger I, Sievers M, Schmitz G (2009) Proceedings of the 7th Modelica Conference, Como, Italy, September 20–22, 2009. doi: 10.3384/ecp09430072
46. Lai JS, Nelson, DJ (2007) Energy management power converters in hybrid electric and fuel cell vehicles. P IEEE 95:766–777. doi: 10.1109/JPROC.2006.890122
47. Spotnitz R, Franklin J (2003) Abuse behavior of high-power, lithium-ion cells. J Power Sources 113:81–100. doi: 10.1016/S0378-7753(02)00488-3

Chapter 6
Lithium-ion Cells in Hybrid Systems

Malgorzata K. Gulbinska, Arthur Dobley, Joseph S. Gnanaraj and Frank J. Puglia

Abstract Despite the versatility of lithium-ion cell chemistry, some specific applications require performance parameters that cannot be met by lithium-ion cells alone. Therefore, the field of hybrid systems containing a lithium-ion component and another power or energy component is rapidly growing. Hybrid power systems such as lithium-ion cells with ultracapacitors, as well as lithium-ion cells with lithium-air batteries are discussed herein.

6.1 Hybrid Systems: General Introduction

The need for power sources that are hybrids of different types of energy storage devices can be easily understood when looking at a Ragone plot (Fig. 6.1), where specific energy (Wh kg^{-1}) is shown as a function of specific power (W kg^{-1}), allowing for a practical comparison of several energy storage technologies at a glance. Certain applications require energy or power levels that exceed those available from a given technology, e.g., high power pulses or prolonged deep discharge periods (high energy) are required periodically while the rest of the time the device must operate within the "standard" metrics range. It follows that the application requirements drive the selection of the hybrid components.

M.K. Gulbinska · A. Dobley (✉) · J.S. Gnanaraj · F.J. Puglia
Yardney Technical Products, Inc., 2000 South County Trail,
East Greenwich, RI 02818-1530, USA
e-mail: adobley@yardney.com

M.K. Gulbinska (ed.), *Lithium-ion Battery Materials and Engineering*,
Green Energy and Technology, DOI 10.1007/978-1-4471-6548-4_6

In this chapter, hybrid powers sources that combine lithium-ion cells[1] with other energy storage technologies are discussed. In particular, three types of hybrid systems are described:

(1) Lithium-ion/Ultracapacitor hybrid that target applications which require sporadic-to-frequent high rate pulses.
(2) Lithium-ion/Lithium-air hybrid system that addresses the application's need for enhanced energy.
(3) Lithium-ion/Fuel cell hybrid where battery is the "high power" component; the optimization of the lithium-ion cell and the selection of the most suitable fuel cell chemistry are discussed.

In most cases, successful hybridization of two or more power sources is expected to prolong the cycle life of a resulting system since each component performs at the power and/or energy conditions that are close to their respective optimum range. The best utilization of a hybrid system is often assured by the sophisticated hybrid electronic battery management system (BMS); much work has been done on simplifying the BMS for the emerging hybrid power systems [1, 11–15]. Even with the current improvements, the need for a DC/DC converter to link the hybrid components to the power bus still contributes to mass, value, life, and cost analysis of the hybrid system and presents an opportunity for improvement.

6.2 Lithium-ion/Ultracapacitor Hybrid System

An ultracapacitor (UCAP), also known as supercapacitor, electrochemical double-layer capacitor (EDLC) or the electrochemical capacitor (EC) is an energy storage device whose specific energy range falls between a battery and a traditional capacitor (Fig. 6.1). EDLCs can be charged and discharged as quickly as capacitors, but possess much greater capacitance. Ultracapacitors perform in a complementary way to lithium-ion batteries, with extremely high power densities but comparatively low energy density. The Ragone plot shown in Fig. 6.1 demonstrates the general capabilities of a lithium-ion battery and an ultracapacitor.

As many new applications, both military and commercial, call for increased energy and power values, it has been of interest to develop an energy storage device that combines the energy capability of a Li-ion battery with the power capability of an ultracapacitor. One potential solution, which has been reported to date, is a hybrid system which features a battery and an ultracapacitor as well as sophisticated, custom electronics to control energy flow between components [1]. The main benefit of this solution is the ability to optimize each component for energy or power.

[1] Energy and power ratings for lithium-ion cells shown in Fig. 6.1 represent the approximated metrics commonly available in commercial devices; specialized high-end lithium-ion units may deliver performance that exceeds the usually reported values.

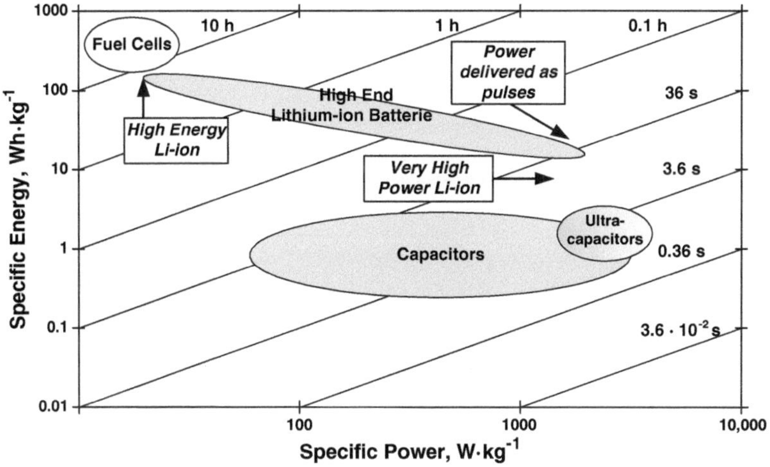

Fig. 6.1 A simplified Ragone plot for selected power sources

Since capacitors exhibit flatter voltage versus time profiles than Li-ion batteries, especially at low temperatures (≤ -20 °C), the benefit of a hybrid system is more significant at low temperatures. This temperature performance differential is even more extreme during the re-charge period of the hybrid system. The downsides of the hybridization approach include the cost of packaging and electronics, the difficulty in sizing the battery and ultracapacitor for a range of usage profiles and the cycle life, which would be ultimately limited by the battery component.

A more advanced solution would require a better integration of capabilities of the battery and the ultracapacitor. Such single, integrated hybrid electrical device has been proposed and built using an advanced lithium-ion battery from Yardney Technical Products, Inc. and high performance ultracapacitors from Maxwell Technologies, Inc. for the National Aeronautics and Space Administration's (NASA) application.

The main benefit from building a Li-ion/UCAP battery hybrid system is that the Li-ion battery in such hybrid can be built to deliver much higher energy density because the burden of delivering high power pulses falls to the ultracapacitor. As discussed earlier in this book, optimizing the construction of a high-end lithium-ion cell usually involves sacrificing power while boosting the energy, or maximizing power at the cost of energy.

An excellent example of this power/energy conundrum can be seen in Yardney's NCP7-type aerospace cell test bed. The specific energy for the most common version of these cells, the Low Earth Orbit (LEO) satellite cell, is 108.3 Wh kg^{-1}. This cell configuration is designed to support moderate loads at relatively low temperatures; supporting current rates up to C/2 at -20 °C. When built in a high rate configuration, the NCP7-type cell-specific energy is only 81.7 Wh kg^{-1}. Yardney has also developed a high energy NCP7 cell configuration that

Table 6.1 Yardney's NCP7 cell configurations

	Normalized cell mass (%)	Specific cell energy (Wh/kg)
High power NCP7	107	81.7
Typical LEO NCP7	100	108.3
High energy NCP7	103	141.7

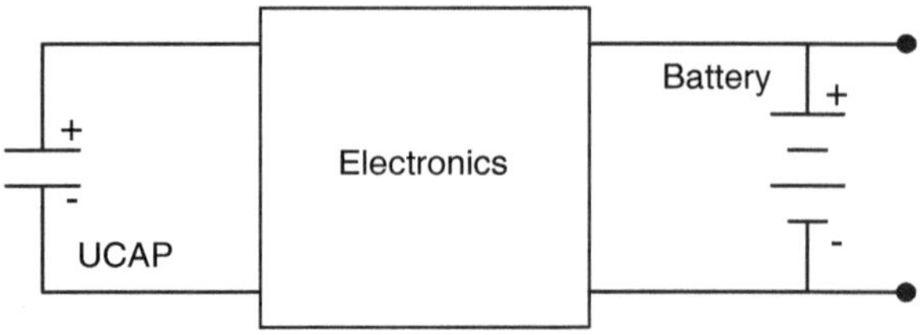

Fig. 6.2 The schematic of the Li-ion/UCAP hybrid battery

has been built into this cell format for qualification testing purposes. The high energy cell delivers 141.7 Wh kg^{-1}. These three versions of the NCP7 cell type differ very little in their mass (Table 6.1) and their volume is the same across all versions but their specific energy differs significantly. The high power version of the NCP7 family pays a specific energy penalty in exchange for the significant increase in power; with continuous discharge current rates as high as 200 °C being handled repeatedly by cells with this configuration.

In a Li-ion/UCAP battery hybrid, there is no need to optimize the lithium-ion cell for power, and therefore, the energy of the cell can be maximized while the high power loads are being handled by the ultracapacitor. Furthermore, since the lithium-ion battery is not being forced to perform at the power extremes, the cycle is expected to be extended as a result of hybridization.

6.2.1 Li-ion/Ultracapacitor Hybrid Construction

The prototype of the Li-ion/UCAP hybrid battery system discussed here was composed of a lithium-ion cell, an ultracapacitor and the commercial-off-the-shelf (COTS) electronics (Fig. 6.2). The system requirements were based on the NASA's Lunar Rover application profile.

The minimum capacity requirement for the Lunar Rover application was no less than 34 Ah and the required discharge power capability was at least 27 kW. In a non-hybridized system, entirely based on lithium-ion modules, the following components would be necessary to fulfill the mission requirements:

(1) Two battery modules, connected in parallel, delivering the nominal capacity of 120 Ah with the nominal voltage of 288 V.
(2) Lithium-ion BMS electronics.

When designing the Li-ion/UCAP hybrid, 60 % depth of discharge (DOD) was assumed for the battery, along with the maximum charge current rate of 0.33 C and the maximum discharge current rate of 0.8 C. Certain aging factors, such as the decrease in capacity and self-discharge were taken into account when sizing the design of the hybrid system capable of functioning through the end of life of both battery and ultracapacitor. The resulting system comprised of:

(1) One 280 V, 60 Ah lithium-ion battery (Yardney Technical Products, Inc.)
(2) One 216 V, 38 F ultracapacitor module (Maxwell Technologies, Inc.)
(3) Hybrid system BMS electronics.

Mass savings approaching 36 % and volume savings exceeding 40 % were achieved in the hybrid system design compared to the battery system design based solely on lithium-ion modules. The mass and volume calculations did not include BMS electronics' mass or volume but it is reasonable to assume that the COTS electronics in the more mature prototype can be selected with a higher priority given to their size and mass optimization.

The initial proof-of-concept hybrid system had the reduced size of 1/300 energy of the full size system and 1/72 power compared to the full size system, discussed above and consisted of the 2 Ah, 32 V prototype lithium-ion battery and two ultracapacitors (52 F, 16 V) connected in series. The testing results presented in the following section refer to the reduced size hybrid system.

6.2.2 Li-ion/Ultracapacitor Hybrid Testing

The hybrid system was charged using the following procedure:

(1) Constant current charge at 2 A (1 C rate) to 32.8 V at room temperature.
(2) Constant voltage charge until the current is less than 0.2 A (or 0.4 A with system) at room temperature.
(3) For the tests at 0, −10 and −20 °C, the system maintained charge at 0.3 A level and was held at the test temperature for at least 1 h before the discharge

The system was discharged according to the following guidelines:

(1) The following discharge power profile was done at four different temperatures, 25, 0, −10, and −20 °C:
    - 504 W pulse (18 A maximum) for 5 s, which corresponds to 9 C pulse current rate
    - 10 W pulse (0.6 A maximum) for 60 s, which corresponds to 0.3 C discharge current rate.

The over-current limit was set at 18 A and the discharge was stopped when the battery voltage lower than 24 V. Cycle life tests were conducted using cycles consisting of the same charge and discharge procedures as listed above, repeated extensively at 25 °C.

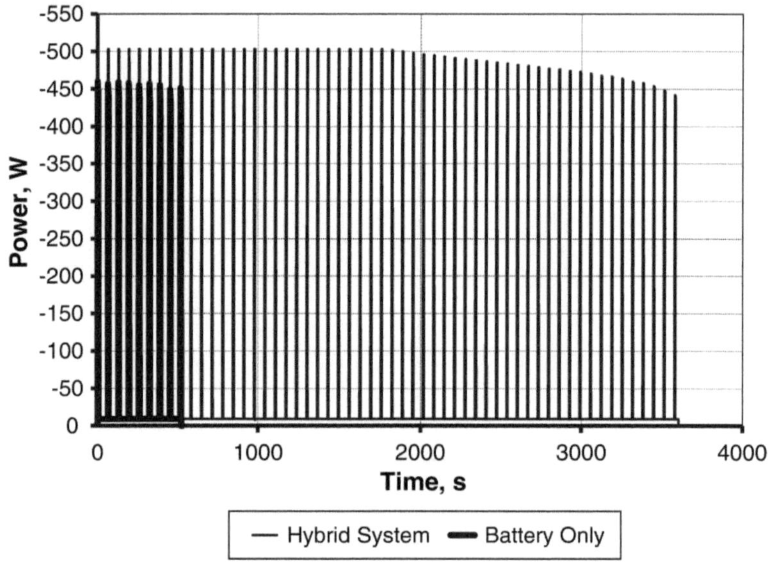

Fig. 6.3 Hybrid system power at 25 °C tests showing cycle life improvement over battery-only option

6.2.3 Li-ion/Ultracapacitor Hybrid Results

At room temperature, without ultracapacitor modules, the battery can run for 585 s (9 cycles) versus 3,516 s (54 cycles) in a hybrid system. The tests showed that the run time of a fully charged battery was increased by 500 % via hybridization, as seen in Figs. 6.3 and 6.4.

When tested at several low temperature conditions, the battery-only system, (optimized for energy) was not able to deliver the required power at temperatures at or below 0 °C. The hybrid system, on the other hand, delivered the power pulses at lower temperatures. The decreased run time was observed due to the increased internal resistance of the battery, aggravated to a greater extent with low temperature conditions ranging from 0 °C, through −10, to −20 °C. The testing results of hybrid system cycling at various temperature conditions are shown in Figs. 6.5 and 6.6.

6.2.4 Li-ion/Ultracapacitor Hybrid Conclusions and Applications

The Li-ion/UCAP hybrid system was successfully built and tested, and has demonstrated 500 % improvement in device run time under 9 C current rate pulse conditions and at low temperatures compared to the lithium-ion battery-only module that had the equivalent nominal energy. The successful implementation of such

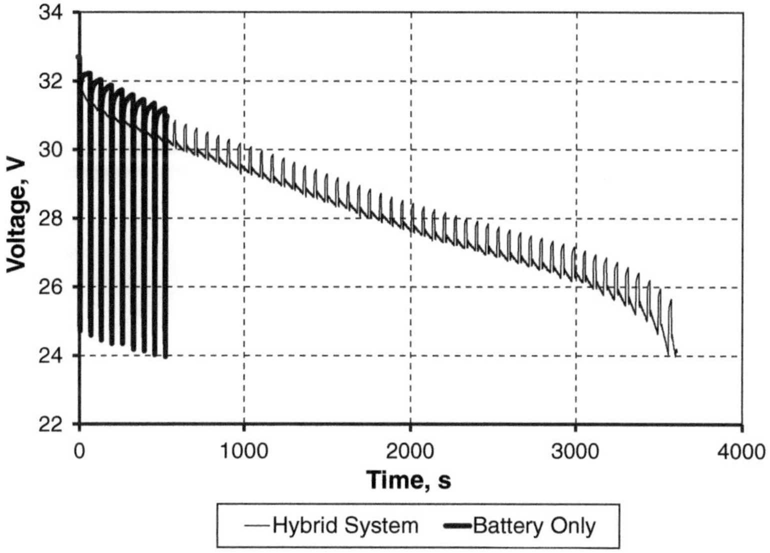

Fig. 6.4 Hybrid system voltage at 25 °C tests showing the improved, narrower operating voltage range for hybrid system

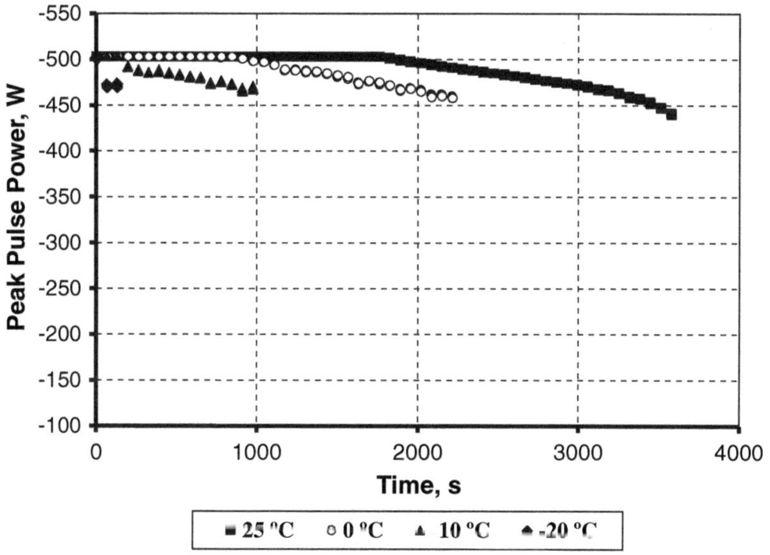

Fig. 6.5 Hybrid system power at different temperature conditions

Li-ion/UCAP hybrid systems with high power and long life would benefit multiple military and commercial communication satellite applications as well as selected Hybrid Electric Vehicle (HEV) and Electric Vehicle (EV) applications.

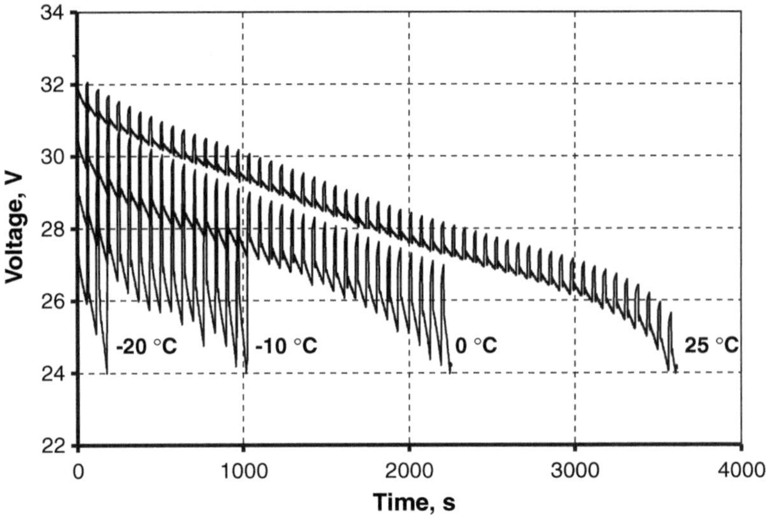

Fig. 6.6 Hybrid system voltage at different temperature conditions

6.3 Li-ion/Lithium-Air Hybrid System

One way to increase the performance of a lithium-ion battery is to apply it in a hybrid system with a metal-air battery. Metal-air batteries can provide a high specific energy to the system (>1,000 Wh kg^{-1}) [2]. The hybrid system is a lithium-ion battery connected to a metal-air battery with electronics. The strategy is that the lithium-ion battery would provide the high currents, and the metal-air battery would provide the low currents. Some systems will have the metal-air battery recharge the lithium-ion battery in addition to providing power for the small electrical loads. The metal-air battery selected for this system is the lithium-air battery. Lithium-air chemistry was selected for its very high theoretical specific energy of 13 kWh kg^{-1}. Practical energy densities are substantially lower now due to the early stage of work on this new technology. Even though the lithium-air technology is new, it was still able to increase the performance of the hybrid system. The lithium-air battery has received much attention due to its very high specific energy, which is the highest for metal air battery systems. The lithium metal anodes are electrochemically coupled to oxygen (from the air) through an air cathode. Oxygen gas, entering the battery through the air cathode, is basically an unlimited cathode reactant source. The two discharge reactions of the lithium-air battery are

$$2Li + O_2 \rightarrow Li_2O_2 \quad E^0 = 3.10\ V \tag{6.1}$$

$$4Li + O_2 \rightarrow 2Li_2O \quad E^0 = 2.91\ V \tag{6.2}$$

Theoretically, with limitless oxygen from the air to react within the air cathode, the capacity of the battery is limited by the lithium anode. Hybrid battery systems have also received much attention lately due to increased performance.

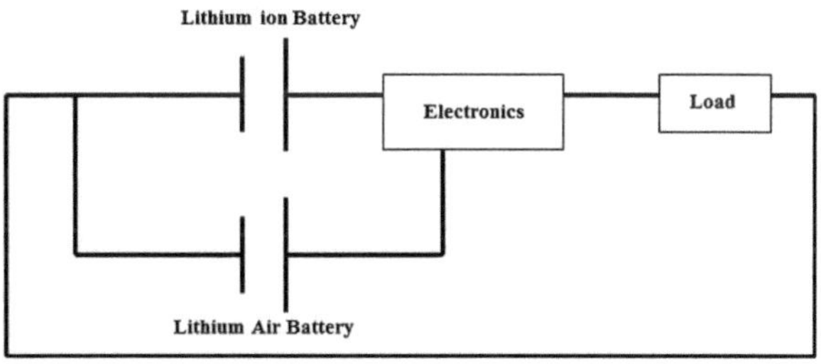

Fig. 6.7 The schematic of the lithium-ion/lithium-air hybrid battery

Fig. 6.8 Photograph of a lithium-air cell, *left* the air cathode, *right* (courtesy of Yardney Technical Products, Inc.)

6.3.1 Li-ion/Lithium-Air Hybrid Battery Construction

The "high-energy" hybrid battery system was composed of a lithium-ion cell, lithium-air cells and simple electronics (Fig. 6.7). Initially, lithium-ion coin cells were used in the hybrid system to prove feasibility. Then lithium-ion pouch cells were used in the hybrid system. Lithium-ion pouch cells (about 120 mAh) were built with a carbon anode, $LiPF_6$ salt in organic solvents was used as electrolyte, and a transition metal oxide cathode was used. Each system had one lithium-ion cell, and two lithium-air cells in series. The lithium-air cell was constructed with a plastic case and involved layering the cell components within it (Fig. 6.8). A lithium metal anode, separator, electrolyte, and catalyzed carbon air cathode were placed inside the medium-sized plastic case. The electrodes and separator were stacked together and placed between two pieces of the case that had a window to allow oxygen to enter. The anode was lithium metal, the electrolyte was 1 M $LiPF_6$ salt in organic carbonates, and the air cathode (about 15 cm^2)

was a laminated structure with a catalyst. The non-aqueous electrolyte was used to increase the safety of the cells compared to the aqueous lithium-air cell, and increase the overall safety of the hybrid system. The system had inexpensive and simple electronic components to allow the cells to work together. Three different electronic configurations were investigated. One configuration was selected due to performance, cost, and simplicity. The hybrid batteries, and the individual cells themselves, underwent testing to investigate discharge performance.

6.3.2 Li-ion/Lithium-Air Hybrid Testing

The hybrid batteries were discharged in oxygen gas, at 20 °C, at about one atmosphere of pressure. The hybrid batteries were discharged in test bags filled with oxygen gas. The bag is sealed and gas tight. The bag can be opened at one end, and cells can be inserted or removed. A valve on the bag allows gas filling and evacuation. A wire feed-through allows leads to connect to the cells and still maintain a gas tight seal. The bags are reusable, and may be used many times. The batteries and cells are placed inside the bag, connected to the wires, then flushed and filled with oxygen. The bags were slightly filled with oxygen gas so the pressure of oxygen inside the bag is about one atmosphere. The hybrid-battery system composed of two lithium-air cells and a lithium-ion cell assembled together for a 6 volt system. Then they were continuously discharged between two currents of 2 and 20 mA. The 2 mA current lasted for 90 s, which was followed by a 20 mA current for 10 s. This was repeated until the lower cutoff voltage was reached, typically 3 V. Two hours rests were included both before and after the discharge for observation. Both a custom made discharge unit and a Maccor unit were used to test the batteries. Since the lithium-air cells are primary cells, new cells were used in each of the tests. The lithium-ion cells are rechargeable and were reused frequently. This constant discharge with pulses was used to simulate electronic and communication equipment.

6.3.3 Li-ion/Lithium-Air Hybrid Testing Results

This hybrid system delivered more energy than adding the energy of each cell type (Table 6.2). The increase in energy was about 130 % for the pulse sequence when compared to sum of the parts. The two types of cells working together provided more energy than the combined efforts of each type of cell. The battery was pulsed many times before reaching the selected lower voltage cut off. Similar values were observed for additional tests, and also tests using coin cells. Additional tests included 12 volt systems and larger capacity cells for both the lithium-ion and lithium-air cells. Other pulse sequences and other environments are expected to increase or decrease the total energy of the discharge.

Table 6.2 Discharge results of individual components and the hybrid battery

Power source	Duration (h)
Li-ion pouch cell alone	31
2 Li-air cells in series	19
Sum of above	51
Li-ion/Li-air Hybrid	69

6.3.4 Li-ion/Lithium-Air Hybrid Conclusions and Applications

The goal of developing a hybrid lithium-ion/lithium-air battery was achieved and demonstrated in the discharge of these hybrid batteries. The combination of lithium-air cells and lithium-ion cells with electronics increased the amount of total energy available. The best designs show the hybrid system delivering twice the energy of the components when run alone. 6 and 12 V hybrid systems were built and tested. Optimization goals begin with the low cost and simple electronics package, and extend onto the overall weight reduction. These hybrid lithium-air batteries promise to provide a high energy density system capable of high rates of discharge. The specific capacity goal of full hybrid batteries is 2,000 Wh/kg. These batteries have the potential to power electronic equipment, vehicles, robots, sensors, munitions, camping equipment, or any equipment where air or oxygen is present.

6.4 Design Considerations for a Fuel Cell/Li-ion Rechargeable Battery Hybrid Power System

Fuel cells are electrochemical devices that convert chemical energy contained in fuel *directly* into electrical energy through electrochemical reactions. Fuel cells consist of an anode, where the fuel is oxidized, a cathode where the oxidant is reduced and an electrolyte which separates anode from cathode and conducts ions. The general classification of fuel cells is usually based on the type of electrolyte used, and their operation conditions are typically related to the characteristics of the electrolyte. More detailed discussion of fuel cells as stand-alone power sources can be found in the next chapter of this book.

Fuel cells are attractive for energy storage applications due to their high power density and low emissions [3]. However, most fuel cell types have issues working independently as a power system. A rapid voltage drop and a slow response to the load demand are the critical issues plaguing fuel cells. Additionally, the efficiency of a fuel cell is a strong function of electrode power density and significant fuel savings are possible when the fuel cell can be tuned to a constant load (e.g., in a hybrid power system with a battery). Thus, in most applications, a battery capable of fast charge and discharge, as well as the power regulation devices, such as DC/DC converters, are required to work with the fuel cell to form a power system to provide regulated, fast-response power to a variable load.

Specifically, in a practical design of the fuel cell/Li-ion battery hybrid, the secondary Li-ion battery acts as the highly responsive power source, while the latent but substantial energy from the fuel cell recharges the battery to sustain an overall system capable of performing the required mission. Ultimately, the fuel cell would be an integral part of a hybrid system to be fully viable. In the effective hybrid system, a high power, Li-ion battery would provide backup power, start-up power, and burst power for a fuel cell-centered system. Further, limiting the peak load on the fuel cell would save significant mass by limiting the fuel cell size and allowing it to operate at its peak efficiency.

Another advantage of the hybrid system is that it allows the fuel cell component to be turned off to conserve energy when power demand is very low. The battery handles the power loads during these hibernation periods and can also be used to reheat the fuel cell to operating temperature when it is needed again.

6.4.1 A Case Study: The Fuel Cell Component Selection for the Hybrid System

The selection of fuel cells for hybridization is, not surprisingly, driven by the specific application requirements. The top-level selection of fuel cell candidates for hybridization is usually governed by the following requirements:

(1) Total system energy requirement
(2) Average system power
(3) Peak power
(4) The required operating temperature range
(5) Cost of the system; very important in the consumer applications, of secondary importance in the military applications.

Following the top-level selection, a detailed survey of a narrowed-down group of fuel cell types can be done. An example of the detailed survey of various fuel cells considered for a hybrid power system is given below, along with the crucial application requirements. Included in the survey were the following candidates: methanol and hydrogen fueled polymer electrolyte fuel cells (PEM; an alternative common abbreviation—PEFC), hydrogen fueled phosphoric acid fuel cells (PAFC) and hydrogen and sodium borohydride ($NaBH_4$) fueled alkaline fuel cells (AFC). The solid oxide fuel cells (SOFC) were excluded from this particular study at the top-level selection stage due to the specific operating temperature requirements. Normally, SOFCs are a very viable option for hybrid power systems.

Two sets of specific operation parameters (two potential unit designs) were identified for an air-independent, low temperature (<100 °C) naval application:

(1) 85 kWh total energy requirement, 8.5 kW average power, 17 kW peak power
(2) 1.2 M Wh total energy requirement, 60 kW average power, and 120 kW peak power requirements.

The comparison points for candidates were as follows:

(1) Fuel storage
(2) Operating limitations
(3) Estimated specific energy, energy density and power density in the particular hybrid system
(4) Technology readiness
(5) Other points such as efficiency and thermal losses and fuel purity requirements

These selection factors are discussed in detail in the following sections.

6.4.1.1 Fuel Storage

Fuel cells are usually open systems, and their energy density and specific energy is largely based on the storage of fuel (and oxidant in the case of air-independent systems). Of the fuel cell types considered here, hydrogen PEMs, PAFCs and hydrogen AFCs use hydrogen as fuel, direct methanol fuel cells (DMFC) use aqueous methanol solution or pure methanol, and (alkaline) direct borohydride fuel cells (DBFC) use sodium borohydride solution as the liquid fuel.

Hydrogen fuel can be stored in four ways: compressed gas, cryogenic liquid, metal hydride, and chemical hydride. In the molecular form (compressed or cryogenic liquid), hydrogen, due to its small molecule size relative to other compounds, has a high leak rate, and requires special tanks to hold its contents. In its compressed gas form, laboratory settings use steel cylinders with internal pressures reaching 200 bar and the large scale storage systems use modified cylinders that can safely hold 300 bar, while having a maximum breaking pressure of 1,200 bar. Hydrogen storage tanks of up to 700 bars (10,000 Ψ) for automotive applications have started being approved. As a cryogenic liquid, hydrogen is cooled to 22 K with pressures around 3 bar. Despite the need for significant layers of insulation, liquid hydrogen tanks hold large quantities of fuel, while being a safer alternative during tank rupture. Hydride storage is promising for low rate applications but is limited by the weight percent of hydrogen stored. At present, the best materials store only 4 % Hydrogen by weight, reducing the practical energy density of the fuel cell system. The main advantage of using hydrogen as fuel is that it is "pure power" that produces only waste heat and water.

Methanol as fuel exhibits less explosive nature than compressed hydrogen. However, when stored as a liquid, methanol can be just as dangerous due to the high flammability, as well as the poisonous character and the absence of taste. During storage, methanol cannot be held in regular steel tanks with rubber gaskets, due to corrosion of rubber. The tank can be made either of stainless steel or glass, but alternative gaskets would need to be found that can withstand methanol and the marine environment. The gasket corrosion issue could be alleviated with regular replacement of the gaskets.

Sodium borohydride ($NaBH_4$, SB) is relatively safe to transport compared to hydrogen or methanol, but is the most expensive of these three fuels. Stored as

Table 6.3 Fuel storage issues for selected fuel cell types [4]

Fuel	Fuel cell type	Energy density (Wh L^{-1})	Specific energy (Wh kg^{-1})	Fuel storage issues
Hydrogen, 25 g L^{-1} (compressed gas)	H_2 PEM fuel cell PAFC H_2 AFC	833.33	1500	High pressure tanks
Hydrogen, 35 g L^{-1} (cryogenic liquid)		1,167	2,167	30 % of energy is lost in the process of liquefaction
Hydrogen, 30 g L^{-1} (chemical hydride)		1,000	1,083	
Hydrogen, 30 g L^{-1} (metal hydride)		1,000	833.33	
Methanol (pure)	DMFC	1,382	1,750	Poisonous, low conversion efficiency
Sodium borohydride	(alkaline) DBFC	1,514	1,403	Developmental

a pure solid or a concentrated liquid solution, SB releases hydrogen gas in the presence of a catalyst, and only has a half-life of about 2 years. SB is used as a hydrogen storage medium, in which an external reactor is used to generate hydrogen through exothermic hydrolysis (Eq. 6.3). There are fuel cell prototypes in development that directly utilize the SB fuel (Table 6.3).

$$NaBH_4 + 2\,H_2O \rightarrow NaBO_2 + 4\,H_2 \tag{6.3}$$

6.4.1.2 Operational Limitations of the Candidate Fuel Cells

In PEM fuel cells, the main operating limitation is the water management. Water is the main product of the cathode reaction, and is evacuated out through the cathode flow channels; more details on PEM fuel cell operation and schematics can be found in the next chapter of this book. Failure to evacuate enough water causes the H_2O condensation within the fuel cell which blocks the reactant flow channels. Removing water excessively, on the other hand, reduces the polymer electrolyte's ability to conduct protons.

Undiluted methanol in DMFCs is subject to significant crossover from anode (fuel) side to the cathode, which results in large amounts of unreacted fuel and energy inefficiency of the DMFC cell. The fuel crossover also causes the very low power density of DMFCs. Using the diluted solutions of methanol in water as fuel in DMFCs significantly diminishes the crossover but decreases the DMFC cell's energy density. In some high-end (mostly military) applications, undiluted methanol is still preferred despite the aggravated crossover issue and this is the case where the high energy density of the DMFC is more important than the energy efficiency of the cell. Because of the above limitations, DMFCs are typically limited to low power applications. In higher power applications, indirect methanol fuel cells with a reformer are preferred.

AFCs often run at a higher temperature conditions than PEM fuel cells, and the typical AFC operating temperatures range from 60 to 200 °C, usually at elevated pressure. Coupled with a liquid electrolyte, elevated temperatures and pressure conditions cause the sealing issues, particularly in the systems, where electrolyte is circulated. On the other hand, the immobilized electrolyte systems are extremely prone to CO_2 poisoning; as the potassium hydroxide (KOH) electrolyte solution reacts with CO_2 (the circulated electrolyte AFCs also show the CO_2 sensitivity but to a lesser extent). The fact that AFCs cannot be exposed to carbon dioxide or carbon monoxide limits their usefulness to CO_2-free environments such as space or underwater applications.

PAFCs run at temperatures around 200 °C. Unlike PEM fuel cells, the electrolyte in PAFCs does not rely on water for proton conduction; therefore water management is not a concern. PAFC systems usually are used in coupled heat and power (CHP) applications for improved efficiency. Due to corrosion caused by phosphoric acid-based electrolyte and dissolution/corrosion of Pt-based catalyst in PAFCs, the operating voltage conditions of PAFCs are limited to less than 0.8 V per cell. The dissolution/corrosion problems also eliminate the option of hot idling of PAFCs. Borohydride energy density at 60 °C would allow a smaller, more adaptive system.

6.4.1.3 Estimated Specific Energy, Energy Density, and Power Density in the Particular Hybrid System

See Tables 6.4 and 6.5.

Table 6.4 Selected performance parameters of fuel cell candidates for the two hybrid systems in consideration

		System 185 kWh total energy, 17 kW peak power		System 2 1.2 M Wh total energy, 120 kW peak power	
Fuel Cell Type	Efficiency	Energy density ($kWh\ L^{-1}$)	Specific Energy ($kWh\ kg^{-1}$)	Energy Density ($kWh\ L^{-1}$)	Specific Energy ($kWh\ kg^{-1}$)
PEM fuel cell	0.5	454.2	796	510.8	917.8
DMFC (pure methanol)	~0.6	109.4	147.4	195.8	259.1
AFC	0.6	268.3	361.1	388.0	565.3

Table 6.5 Power density for selected fuel cell candidates

Fuel cell type	Power density ($mW\ cm^{-2}$)
PEM fuel cell	>800
DMFC (pure methanol)	100
AFC	3,400
PAFC	200

6.4.1.4 Comparative Technology Readiness

Fuel cells have been used in applications for decades. PEMs are able to produce high current densities while being able to start and react quickly. Current average lifetime is around 8,000 h [5], but some companies claim longer life spans, for example, 15,000 h claimed by Gore Fuel Cell Technologies [6]. While the fuel cell technology is well demonstrated, development efforts are focused on improving durability and lowering the cost.

DMFCs are usually regarded as the fuel cell type that is closest to commercial viability, despite having a relatively high cost of entry into the market. The most fitting application for DMFCs is the commercial market of portable electronics, such as mobile phones and laptop computers [7].

AFCs were used on the 1960 Apollo spacecraft [8], but because of their intolerance to carbon dioxide, their scope has been limited to places where pure hydrogen and oxygen are needed. The expected lifetime of AFCs is between 2,600 and 5,000 h, but AFCs have strong reliability (The Space Shuttle Orbiter produces electricity from an AFC with no backup power supply).

PAFCs have been available for decades, producing a maximum 5 MW of power for a large stationary cell stack. Current models, produced by UTC Power, have a lifetime of 40,000 h. Multiple industrial companies including UTC Power (US) have been developing PAFCs, but mostly for stationary power generation; the most current interests are to increase the lifetime to 80,000 h.

6.4.1.5 Other: Efficiency, Thermal Losses, and Fuel Purity Requirements

Thermal Losses. The relatively low energy efficiency of the fuel cells, as shown in Table 6.4, is the cause that maintaining the stealth of a submersible for fuel cell-containing power sources is extremely challenging. At a 17 kW output the vessel will need to dissipate at least a comparable amount of heat, possibly more. By contrast, a Li-ion battery operates at over 96 % energy efficiency and a silver zinc battery operates at about 85 % efficiency. Since all of the "waste" energy is ultimately converted to heat, fuel cells would be producing significantly larger heat signature than a battery. The thermal detection impact would be highly dependent on depths and currents and needs to be considered on an application and mission-specific basis.

Since the reactions in all fuels are exothermic, the cells would need to be cooled. Polymer electrolyte fuel cells use bipolar plates in which cooling channels are embedded; coolant water can be run through these channels to provide the appropriate cooling [9]. Temperature changes due to variations in the applied load are generally gradual, allowing for proper thermal management when loads are increased [10]. PAFC also use water or special synthetic oil running through cooling channels between cells to cool the system. However, since the operating temperature of the PAFCs is elevated, water may need to be pressurized. In (non-immobilized) AFCs, the circulating electrolyte can be used for cooling [9].

Fuel Purity Requirements. Generally, low temperature fuel cells have more stringent fuel quality requirements compared to high temperature fuel cells. Sulfur-containing impurities (e.g., sulfur dioxide—SO_2 and hydrogen sulfide—H_2S) are poisonous to all types of fuel cells.

Specifically, PEMs cannot tolerate CO concentrations above 1 ppm. NH_3 is also known to be a contaminant for PEM fuel cells. There is no established fuel quality requirement for DMFCs, since the methanol used is usually pure enough to avoid any concerns. As mentioned above, AFCs, particularly their immobilized electrolyte type, cannot tolerate any CO/CO_2 neither in fuel nor in the oxidizer. This requires a very stringent clean-up of hydrogen if it is produced from a hydrocarbon fuel. Commonly, PAFCs use hydrogen from reformation of hydrocarbons, which will contain CO/CO_2 and unreacted hydrocarbons in addition to H_2. Typically, CO_2 and unreacted hydrocarbons in PAFCs act as diluents and do not affect the electrochemical reaction. CO is poisonous to the electrochemical reaction, due to absorption of CO in catalyst sites but the tolerance level in PAFCs is considerably higher than in PEMs. NH_3 reacts with the phosphoric acid to form a phosphate salt $(NH_4)H_2PO_4$, therefore NH_3 should be completely removed from the hydrogen fuel prior to entering the PAFC stack. Since N_2 can react with H_2 to form NH_3, the allowable N_2 concentration is also limited in PAFCs. The summary of common types of fuel cells' contaminants and their allowable concentration is shown in Table 6.6.

6.4.2 Power Management and Design Optimization of Fuel Cell/Battery Hybrid System

Most hybrid power systems require sophisticated power management in order to best perform their function and derive the maximum performance from their respective components. Fuel-cell-containing hybrids are no exception and an additional degree of complexity is introduced to the power management circuitry by the wide temperature range of operation for most fuel cells as well as the complex balance of plant (BOP) that comes with all fuel cells. BOP components encompass all the air/gas- and fuel-management devices (compressors, valves, seals, pumps, etc.) as well as the fuel cell stack components (electrodes, bipolar plates, membrane assemblies, etc.). Some of the BOP components require sensing and control of their own (e.g., metered fuel delivery, membrane humidification in PEMs, etc.), which would require secondary integration with the hybrid-level controls.

When designing a hybrid system, there are two general ways of setting up the power sharing and control: passive or active power management [11]. The passive power sharing involves a direct connection between a fuel cell and a battery with the electronic components being kept to a minimum. An example of the passive power sharing in the battery-fuel cell hybrid is the system built by Jones et al. [12], which had a voltage limiter to prevent battery overcharge and a simple circuit board performing the auxiliary functions (operating the cooling fan etc.).

Table 6.6 Common impurities in fuel for various fuel cells [4]

Fuel cell type	Impurity	Allowable concentration
PEM fuel cell	H_2S, SO_2	<4 ppm (total)
	CO	0.2 ppm
	NH_3	0.1 ppm
DMFC	–	–
AFC	CO_2	– (immobilized electrolyte)
PAFC	H_2S, CO	<50 ppm (total)
	CO	0.3 %
	NH_3	–
	N_2	<4 %

A passive hybrid system offers low complexity and higher efficiency compared to an active hybrid but requires precise voltage matching between the fuel cell and the battery, which can be disrupted as the result of aging process concerning either of the hybrid components. Another drawback of the passive hybrid system is the fact that the maximum power output depends on the state-of-charge of the battery, which also can be affected by the aging process.

The active hybrid system incorporates a step-up or step-down controllable DC/DC converter, with the fuel cell connected to the battery via a DC/DC converter while the battery is directly connected to the load [13, 14]. The voltage matching of components is done via DC/DC converter, and any combination of fuel cell and rechargeable battery can be used. For example, the PEM fuel cell can be combined with several other battery types, including lithium-ion battery [13]. The installation of the DC/DC converter usually reduces the total energy efficiency of the active hybrid system but with the proper design and sizing of the components, the efficiency can be brought back to >90 % [15].

Component sizing and the power management strategy are two main factors that affect the performance of the hybrid system and their optimization is usually done through multiple-parameter modeling [16]. The modeling parameters take into the account the individual characteristics of each power source and the requested load profile scenarios and allow for simultaneous optimization of all the hybrid component outputs/size.

6.4.3 Designing the Best Lithium-ion Battery for a Battery/Fuel Cell Hybrid System

Selection of the lithium-ion battery parameters for use in the fuel-cell/battery hybrid system must start from the analysis of the predicted usage profile. The analysis includes estimates of the expected power and energy requirements for the battery within the hybrid system and selecting the most effective cell formulation and/or chemistry for the task. The optimization of the lithium-ion cell characteristics

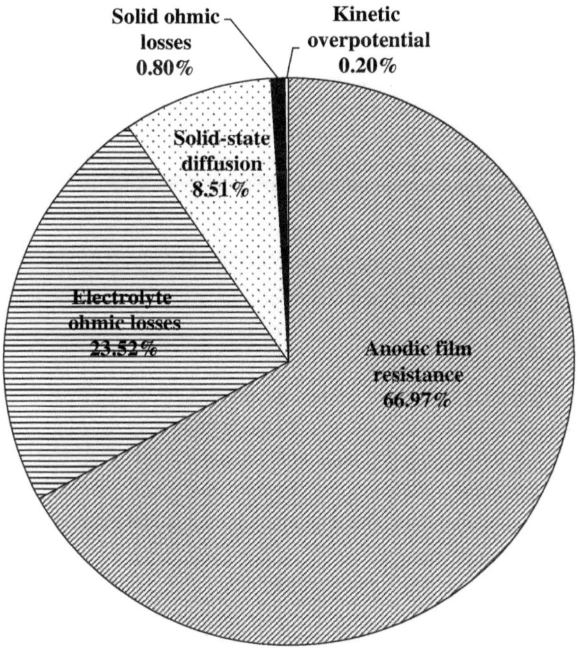

Fig. 6.9 Summary of the inefficiencies in lithium-ion battery model and their origin [17]

for the application in hybrid systems is similar to the stand-alone cell design and the similar methodology can be used in both cases. The stand-alone cell/battery parameters optimization was discussed in the Chap. 3 of this book.

In the fuel cell/Li-ion battery hybrid, the battery generally ends up being the "high power" component and the optimization of the cell proceeds to develop the configuration capable of handling short (minutes or seconds) but relatively large (2 C or more) discharge current pulses. The 2 C discharge current is a practical "limit" for many commercial batteries beyond which the cell would age excessively upon frequent pulse application. For high-end batteries, the 2 C limit is not as firm; certainly, cells' configurations exist that can exceed this value by as much as hundred times. However, the cost of such formulations and the cells' raw materials is currently prohibitive for their mass-production. Still, for the mass-produced "high power" cells, the optimization for power must be done to provide the best-performing hybrid component. In most cases, optimizing battery for power performance means reducing the internal resistance within the cells. Figure 6.9 illustrates the biggest contributors to lithium-ion cell's total internal resistance [17]. The pie chart shown in the Fig. 6.9 illustrates the results of modeling of lithium-ion cell components done by Fellner and Newman [17], where the percentage conductivity losses (i.e., resistance gains) were determined by setting the selected parameter to either a large or a small value to negate its effect in the model.

From the results summarized by chart (Fig. 6.9) it is apparent that carbon-based anodes can contribute as much as two-thirds of the internal cell resistance. This is mostly due to the resistive solid electrolyte interphase (SEI) film formation on the surface of anode's particles. Moreover, the SEI formation can be exacerbated by high surface area of anode material, which in turn is related to the anode's average particle size; with smaller particles often having larger surface area. Thus, when new anode materials are being developed, efforts are being made to keep their surface area at the minimum, usually through utilization of surface coatings and/or treatments. Keeping the anode's surface area low also helps to decrease the undesirable side reactions and the losses in initial cell capacity that is sacrificed when SEI is formed (SEI formation irreversibly consumes lithium-ions). For more discussion of carbonaceous anode materials and the surface area considerations, the reader is directed to Chap. 1 of this book. Despite the potential adverse effect on surface area value, keeping electrode particles as small as possible is important because of the solid state diffusion resistance which is lowered in particles with small diameter. Smaller particles also exhibit lower local current density and thereby decrease ohmic losses within the solid fraction of the electrode. The current density and the discharge duration impact the desired electrode thickness and porosity in a battery candidate for the hybrid power system (Chap. 3). If more power is required for a larger device to be powered, a battery with more electrode area is appropriate.

Another important contributor to the cell's internal resistance is the electrolyte, which comes in at almost one quarter of the total cell's resistance. From the past work on electrolytes' conductivity [18, 19], the salt concentration in electrolyte should be optimized for the maximum conductivity of the electrolyte solution. This is because the short high current pulses during load-leveling of the hybrid should not cause the development of salt concentration gradients in the cell, which can adversely impact cell's performance and life.

6.4.4 Fuel Cell/Li-ion Hybrid System: Conclusions

Fuel cells are a mature enough technology that should be considered for larger underwater vehicles. The mass requirements of fuel storage render the viability of fuel cells questionable for smaller systems, despite the successful use in several small unmanned aerial vehicles (UAVs). The testing of UAVs typically involved very short testing times and thus required limited fuel.

Several fuel cell systems, most notably, AFCs and PEMs, are viable and therefore the specific application would likely drive the final selection. The startup times are still a concern as well as the power capabilities. Both the slow start-up and the insufficient power capability can be addressed by utilizing a battery-fuel cell hybrid. Furthermore, a battery-fuel cell hybrid system has a built-in emergency power backup.

A condenser would likely be required for stealth operation to eliminate the release of water vapor bubbles and to maintain vessel mass and buoyancy. In naval applications, there are large amounts of water available for cooling of the power sources. Fuel storage could be an issue, but would be partly alleviated by the fact that modern submarines produce hydrogen as a waste product in the generation of oxygen and this could become a refueling source, limiting the need for large amounts of hydrogen storage.

6.5 Hybrid Systems: Closing Remarks

In this chapter, three types of hybrid systems were described:

(1) The Li-ion/ultracapacitor hybrid
(2) The Li-ion/Li-air battery system
(3) The Li-ion/fuel cell hybrid

Significant improvements in performance over the individual components were reported for the experimental systems (1) and (2), whereas the discussion of the fuel cell selection considerations and the Li-ion cell optimization process for the hybrid were discussed for the system (3). Multiple military and commercial devices can be prospectively powered by the hybrid systems. The battery/capacitor hybrid consists of the components with the greatest degree of maturity whereas the lithium-ion/fuel cell hybrid has the most parameters and auxiliary equipment that add to the complexity of the system design. However, the Li-ion/fuel cell systems possess the most flexibility in terms of the operating conditions and characteristics and could be tailored to match the applications requirements very closely. In general, hybrid power systems are expected to successfully dominate several niche markets with wide-spanning power and energy requirements. In the hybrid electric vehicles, the hybrids power systems will need to compete with the advanced internal combustion engines and their sizing, cost, and durability are expected to be the main challenges.

6.6 Further Reading

Battery Technologies

(1) Reddy TB, Linden D, eds (2010) Linden's handbook of batteries 4th edition. McGraw-Hill, New York

- This book covers multiple topics on battery fundamentals, chemistry types and current trends; a chapter on metal-air batteries (including Li-air) is also included

Capacitors and Ultracapacitors

(1) Kotz R, Carlen M (2000) Principles and applications of electrochemical capacitors. Electrochim Acta 45:2483–2498. doi:10.1016/S0013-4686(00)00354-6
(2) Burke A (2000) Ultracapacitors: why, how, and where is the technology. J Power Sources 91:37–50. doi:10.1016/S0378-7753(00)00485-7

Fuel Cells

(1) Larminie J, Dicks A (2003) Fuel cells systems explained 2nd edition. John Wiley & Sons, Ltd, Chichester, England

– This reference provides an overview of system level design considerations

(2) O'Hayre RP, Cha SW, Colella W, Printz FB (2006) Fuel Cell Fundamentals. John Wiley & Sons, Inc. Hoboken, New Jersey

Power Sharing and Management in Hybrid Power Systems

(1) Gao L (2005) Power enhancement of an actively controlled battery/utracapacitor hybrid. IEEE Trans. Power Electron 20:236–243. doi:10.1109/TPEL.2004.839784
(2) Kuperman A, Aharon I (2011) Battery – ultracapacitor hybrids for pulsed current loads: a review. Renew Sust Energ Rev 15:981–992. doi:10.1016/j.rser.2010.11.010
(3) Chandrasekarana R, Bia W, Fuller TF (2008) Robust design of battery/fuel cell hybrid systems – methodology for surrogate models of Pt stability and mitigation through system controls. J Power Sources 182:546–557. doi:10.1016/j.jpowsour.2008.04.038

Acknowledgments The Authors of this chapter would like to gratefully acknowledge the support of their work from the following financial sources:

(1) Dr. Henry H. Yoo Civ USAF AFMC AFRL/RVSV, and Dr. David C. Chapman Civ USAF AFMC AFRL/RVSV, SBIR AF Res. Laboratory (DET 8), Contract # FA9453-10-M-0130
(2) Dr. David C. Chapman Civ USAF AFMC AFRL/RVSV, SBIR AF Res. Laboratory (DET 8), Contract # FA9453-11-C-0167
(3) Verdeyen, William K (JSC-ER411), NASA, Contract #NNJ07JB33C
(4) Michael A. Everett, Dr. Kimberly McGrath, Rong Yan, Mikael Setterberg, Ilya Kaminsky, Dr. Priya Bendale, Maxwell Technologies Incorporated 3912 Calle Fortunada San Diego, CA 92123
(5) U.S. Department of Defense, SBIR grant Army Phase II SBIR, "Lithium-Air/Lithium-Ion Hybrid Battery for Military Use", Contract #W15P7T-07-C-H406
(6) Verdeyen, William K (JSC-ER411), NASA, Contract #NNJ07JB33C
(7) U.S. Department of Defense, Navy Contract #N00014-07-M-0110
(8) U.S. Department of Defense, Navy Contract #07PR06126-00
(9) U.S. Department of Defense, Navy Contract #N65536-10-M-0030

References

1. Awerbuch JJ, Sullivan CR (2008) Control of ultracapacitor-battery hybrid power source for vehicular applications. IEEE, Energy 2030 Conference Proceedings 1–7. doi:10.1109/ENERGY.2008.4781003
2. Christensen J et al (2012) A critical review of Li/air batteries. J Electrochem Soc 159:R1–R30. doi:10.1149/2.086202jes
3. O'Hayre RP, Cha SW, Colella W, Printz FB (2006) Fuel Cell Fundamentals. John Wiley & Sons, Inc. Hoboken, New Jersey

4. http://www1.eere.energy.gov/hydrogenandfuelcells/storage/current_technology.html. Accessed 1 May 2013
5. Fuel Cell Handbook, by EG&G Technical Services, Inc. for U.S. Department of Energy, Seventh Edition, 2004
6. Cleghorn S, Kolde J, Liu W (2003) Catalyst-coated composite membranes. In: Vielstich W, Lamm A, Gasteiger HA (eds) Handbook of fuel cells—Fundamentals, technology and applications. John Wiley & Sons, Chichester, UK
7. Brandon NNP, Thompsett D, eds (2005) Fuel Cells Compendium. Elsevier Science, Oxford, United Kingdom
8. Kordesch K et al (2008) The safe and economic revival of alkaline hydrogen/air fuel cells with circulating electrolytes. Recommended for vehicles using battery hybrid systems and hydrogen from ammonia crackers. ECS Trans 11:167–185. doi: 10.11498/1.2992503
9. Blomen LJMJ, Mugerwa MN (1993) Fuel Cell Systems. Plenum, New York
10. Tang Y et al (2010) Experimental investigation of dynamic performance and transient responses of a kW-class PEM fuel cell stack under various load changes. Appl Energy 87:1410–1417. doi:10.1016/j.apenergy.2009.08.047
11. Yufit V, Brandon NP (2011) Development and application of an actively controlled hybrid proton exchange membrane fuel cell-Lithium-ion battery laboratory test-bed based on off-the-shelf components. J Power Sources 196:801–807. doi:10.1016/j.jpowsour.2010.06.029
12. Jones PB, Lakeman JB, Mepsted GO, Moore JM (1999) A hybrid power source for pulse power applications. J Power Sources 80:242–247. doi:10.1016/S0378-7753(99)00074-9
13. Gao LJ, Jiang ZH, Dougal RA (2005) Evaluation of active hybrid fuel cell/battery power sources. IEEE Trans Aerosp Electron Syst 41:346–355.
14. Jiang ZH, Gao LJ, Dougal RA (2007) Adaptive control strategy for active power sharing in hybrid fuel cell/battery power sources. IEEE Trans Energy Convers 22:507–515. doi:10.1109 /TEC.2005.853747
15. Jiang ZH, Dougal RA (2006) A compact digitally controlled fuel cell/battery hybrid power source. IEEE Trans Ind Electron 53:1094–1104. doi:10.1109/TIE.2006.878324
16. Kim MJ, Peng H (2007) Power management and design optimization of fuel cell/battery hybrid vehicles. J Power Sources 165:819–832. doi:10.1016/j.jpowsour.2006.12.038
17. Fellner C, Newman J (2000) High-power batteries for use in hybrid vehicles. J Power Sources, 85:229–236. doi:10.1016/S0378-7753(99)00344-4
18. Xu K (2004) Nonaqueous liquid electrolytes for lithium-based rechargeable batteries. Chem Rev 104:4303–4417. doi:10.1021/cr030203g
19. Blomgren GE (1985) Properties and structures of electrolyte solutions for lithium batteries. J Power Sources 14:39–44. doi:10.1016/0378-7753(85)88008-3

Chapter 7
Competing Technologies Landscape

William A. McPhee

Abstract Lithium-ion cells are one of many options among a multitude of energy storage devices. This chapter discusses some of the most prominent competing energy storage options, organized by the application environment. A socioeconomic commentary is provided for the current and future situation of the selected energy storage market sectors.

7.1 Introduction

Lithium-ion batteries compete against a number of other energy storage and generation devices. The suitability of lithium-ion batteries for a particular application depends on a number of metrics. Convenient top-level application categories are:

(1) Man-portable applications
(2) Transportable applications
(3) Stationary applications

Regardless of the category or end application, price is always a consideration. Although a premium price can be leveraged for aerospace and defense applications, the end users will require a budgetary restraint. For the best price comparison within various technology platforms, it is typically helpful to express all metrics in terms of the cost unit ($) per power (Watts[1]) and energy (Joules[2]) units,

[1] Watts is equivalent to Joules per second.

[2] Note that typically energy in batteries, fuel cells, and other energy devices is expressed as watt-hours (Wh) $\mathrm{Wh} = \mathrm{Ws} \times 3,600 = \mathrm{J s^{-1} s} \times 3,600 = \mathrm{J} \times 3,600$.

W.A. McPhee (✉)
BASF Corporation, 25 Middlesex/Essex Turnpike, Iselin, NJ 08830-0770, USA
e-mail: wagmcphee@gmail.com

M.K. Gulbinska (ed.), *Lithium-ion Battery Materials and Engineering*,
Green Energy and Technology, DOI 10.1007/978-1-4471-6548-4_7

i.e., \$/W or \$/Wh. Quite often for portable and transportable applications, the cost metric is further refined to capture energy density and specific energy, i.e., \$/Wh L^{-1} and \$/Wh kg^{-1}. Comparing various technologies based on these normalized metrics yields meaningful results, as opposed to using the non-standardized power or energy units which do not paint the complete competing technologies' landscape. A majority of the literature reviews summarize and compare various technology platforms using the Ragone plot that has energy density (Wh kg^{-1}) as a function of power density (W kg^{-1}) with bubbles that present various technologies. Once the internal combustion (IC) engines are added to the plot, it is clear that the alternative power sources still require very significant development to be competitive in terms of energy density and power density.

Regarding power density, supercapacitors rank higher than batteries which are in turn rated higher than fuel cells. The reverse trend is seen for energy density and specific energy. Therefore, research on these technologies tends to gravitate toward improvement of the weakest metric. The suitability of a given technology for a particular application is usually determined by the strongest performance metric, with fuel-powered devices (IC engines, fuel cells, and sterling engines) being used for the longest mission profiles while supercapacitors are preferred for short-duration, high-power applications.

Batteries in general are more similar to fuel cells than to supercapacitors in their performance characteristics. Specifically, lithium-ion batteries have the highest energy density and power density of all the mass-manufactured battery types. The fundamental limitation of these technologies at the moment is the inability to deliver both high power density and energy density from the same device. IC engines by comparison have both high energy density and power density. IC engines can also be scaled dramatically from the small version found in model planes and boats to the largest ship diesel engines. Research is also still ongoing on IC engines to render them more economical and efficient. Therefore, a very significant amount of R&D into alternative energy sources is still needed before IC engines are displaced from mass markets.

This chapter is an overview of competitive technologies touching on key issues. The reader is then directed to other sources for a deep dive into the technologies. A summary of references with comments on their focus is provided for the reader at the end of the chapter. The technologies will be compared in terms of their use and performance in the three aforementioned top-level application categories. The strengths and weakness of the technologies will be related to the key metrics already discussed.

7.2 Man-Portable Applications Background

Man-portable applications include the ubiquitous smart phone, power tools, and any other device that can be carried by an average person for a reasonable distance. There are clearly subcategories, since an iPod Nano™ is very different from a military-grade, field-deployable battery charger. In the world of high-tech consumer electronics, user experience is everything. As such, the emphasis is placed on the high performance delivering lithium-ion battery technology, the preferred energy source for commercial portable applications. As the size of the device increases, other technologies become more technically competitive and the cost metric starts to dominate the application requirement. For this review less than 1,000 W and 10,000 Wh will be considered man-portable.

7.2.1 Other Battery Technologies for Portable Applications

In general, other battery technologies are not competitive in portable applications. Lithium-ion chemistry has better power and energy density than other competing battery chemistries. The proliferation of lithium-ion technology has seen the majority of rechargeable battery packs using this chemistry. For example, personal portable audio, video, and phone devices all use batteries with lithium-ion chemistry. Primary batteries are also slowly being replaced by this chemistry. Higher end primary batteries for consumers are lithium-ion chemistry and are typically targeted at more power-intense applications, i.e., digital cameras. The disposable nature of primary batteries means that cost is a dominant metric. Lithium-ion chemistry for primary batteries has been pushed toward a commodity product which has driven down cost. However, the least expensive primary batteries are still alkaline and zinc–carbon chemistries. There are some niche applications in which other batteries' chemistries have superior technical performance that dictate their use. Zinc-air chemistry for use in hearing aid batteries is an example, but lithium-ion chemistry has been introduced by some manufacturers into this market as well. Currently, Audiotoniq, Inc. (Austin, Texas) is marketing a rechargeable cochlear hearing aid based on lithium-ion chemistry. In addition, Siemens Corporation is also developing a range of hearing aids with rechargeable batteries. Until recently, Otologics, LLC (Boulder, Colorado) has been[3] one of three medical device companies to produce middle ear hearing implants with a lithium-ion battery. The other two companies that produce the middle ear hearing implants are Envoy Medical and Med-El Corporation.

7.2.2 Fuel Cells for Portable Applications

Fuel cells are an example of a technology that has been mentioned as the solution to many of humanity's energy needs, but despite voluminous research and development efforts, no mass commercialization was achieved to this date. The

[3] http://www.bizjournals.com/stlouis/blog/BizNext/2012/07/otologics-files-for-bankruptcy.html. Accessed 30 April 2013.

chemistry of fuel cells can be varied, however; typically, the hydrogen–oxygen system is highlighted in the classic literature since the end product is water. This electrochemical reaction can be achieved in a number of ways depending on the operating temperature and electrolyte. In general, a low temperature system has hydrogen protons transported from the anode to the cathode. The exception is the alkaline system in which the hydroxide ions are transported from the cathode to the anode. At an intermediate operating temperature of 600 °C, the dominant fuel cell system is molten carbonate, in which carbonate ions are transported from the cathode to the anode. Although the charge-carrying ion is a carbonate, the fuel is still hydrogen and oxygen with water as the product, since the carbonate ion is recycled back to the cathode as carbon dioxide. At high operating temperatures of >800 °C, oxygen ions are typically transported from cathode to anode. Rudimentary sketches of these systems are found in Fig. 7.1.

The electrochemical reactions in a low-temperature Proton Exchange Membrane (PEM) fuel cell are shown below:

$$\text{Overall, PEM:}\quad H_2 + \frac{1}{2}O_2 \rightarrow H_2O \quad E^\circ = 1.229\ \text{V} \tag{7.1}$$

$$\text{Anode, PEM:}\quad H_2 \rightarrow 2H^+ + 2e^- \tag{7.2}$$

$$\text{Cathode, PEM:}\quad \frac{1}{2}O_2 + 2H^+ + 2e^- \rightarrow H_2O \tag{7.3}$$

The most prominent class of fuel cells for portable applications is the group of PEM fuel cells and their subgroup—Direct Methanol Fuel Cells (DMFC). Their advantages are summarized below:

(1) The PEM fuel cells utilize environmentally friendly fuel—oxygen and hydrogen that can be produced via electrolysis of water. Electrolysis of water when using renewable resources (i.e., solar or wind) provides a clean source of fuel that is then reverted to the water by the fuel cell. Currently however, the majority of the world's hydrogen is generated from hydrocarbon fuels since the economics are more favorable than electrolysis of water. The current high cost of electrolysis is related to the raw material cost of cell hardware and the limited volume of cell hardware within the market. If widely adopted, low pressure electrolysis can be reasonably cheap, if auxiliary systems are utilized for gas cleanup (humidity removal) and gas compression.

(2) The PEM and DMFC systems (Fig. 7.2) have no or a limited number of moving mechanical parts, which is especially advantageous in defense applications. One key advantage of a system with no or a limited number of moving parts is the ability to limit the noise output of the power generation device. Although a pump is typically used for fuel delivery, the pump's noise signal can be easily minimized. In DMFC it is also possible to utilize a wicking mechanism (no mechanical movement) for fuel delivery instead of a pump.

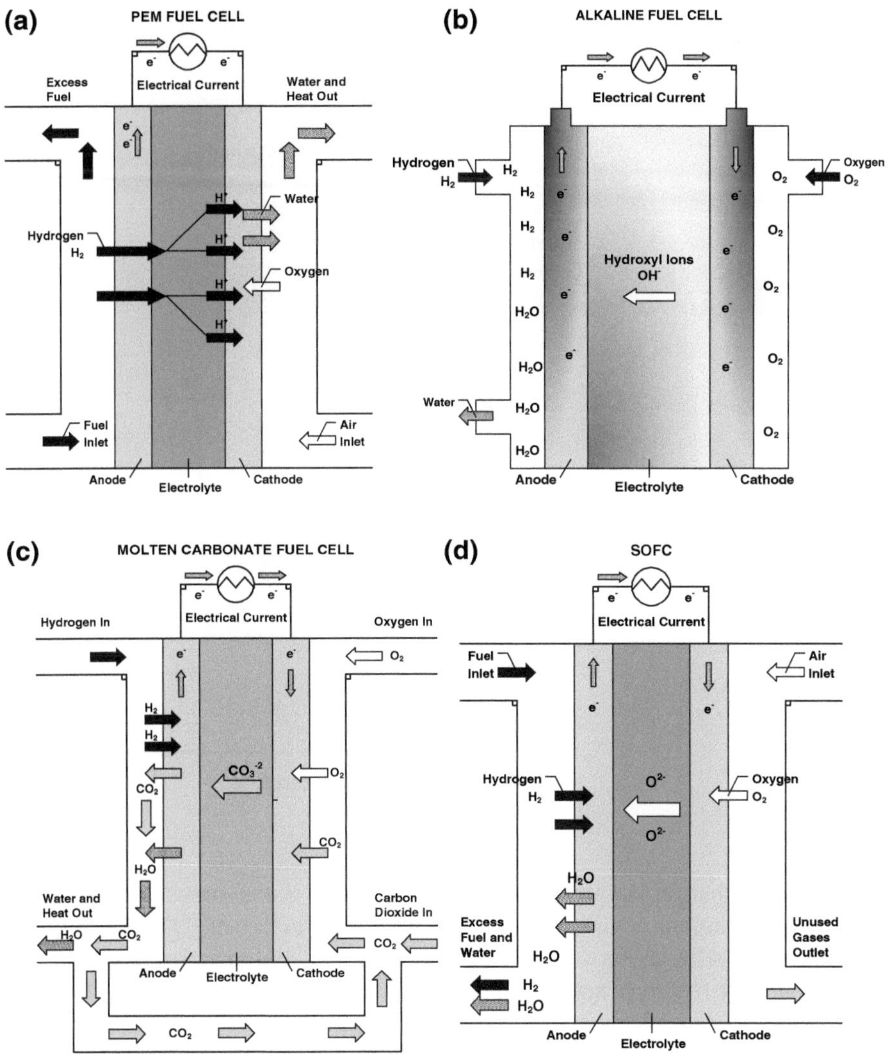

Fig. 7.1 Sketches of a Proton Exchange Membrane (PEM) fuel cell, low temperature (**a**); an Alkaline Fuel Cell (AFC), low temperature (**b**); a Molten Carbonate Fuel Cell (MCFC), intermediate temperature (**c**); a high temperature Solid Oxide Fuel Cell (SOFC) (**d**)

The disadvantages of the PEM and DMFC technologies are listed below:

(1) PEM fuel cells use gaseous fuels but hydrogen storage is difficult to achieve at the energy and power densities sufficient for portable and transportable applications. Oxygen storage is typically not needed since air can be used for the cathode fuel source.

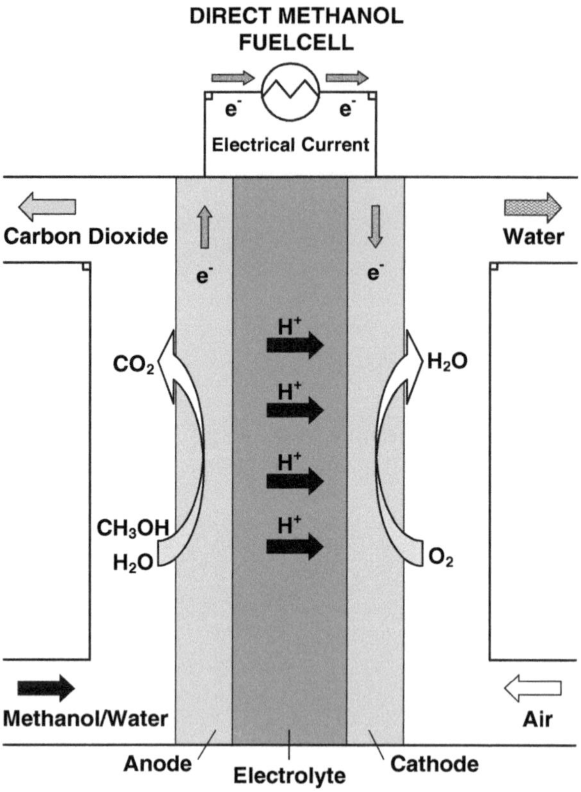

Fig. 7.2 DMFC cell-level schematic

(2) Proton Exchange Membrane (PEM),[4] commonly synonymous with DuPont's Nafion® membrane and used in both PEM fuel cells and DMFCs, requires aqueous conditions for suitable ionic conductivity and therefore needs insulation against freezing, and requires rigorous humidity control. Although other membrane chemistries are available, the sulfonated tetrafluoroethylene-based fluoropolymer-copolymer chemistry used in Nafion® and other brand name membranes is still the material of choice.
(3) Proton conduction in these fuel cells has an undesired side effect of osmotic water drag, also known as the electro-osmotic drag, from anode to cathode which causes the necessity for water recycling from anode back to the cathode at the system level.
(4) PEM cells and DMFCs exhibit poor response to variable loads. Although these fuel cells will operate under loads with a limited degree of variability, they operate best under steady state conditions. As with all chemical and

[4] Note the ambiguous use of the PEM acronym, which in some cases denotes the entire PEM fuel cell or just a PEM membrane.

electrochemical reactions, the kinetics and mass transfer are more favorable if the conditions are steady state. Although variations in operating performance can be accommodated by the chemical reactions, durability of the system is typically directionally proportional to the stability of the system. In hybrid systems there is a distinct advantage gained by pairing a fuel cell (base load) with another power device (peak loads) that allows the power systems to operate in their most reliable states while giving the end user the widest possible range of operating conditions.

(5) PEM and DMFCs rely on catalysts that contain platinum (Pt) group metals. Much research has been done to date on alternative, non-precious metal catalysts, but at less than 300 °C, Pt group metals are required.

(6) Sulfur (S) and carbon monoxide (CO) poisoning of catalysts occur at standard operating conditions of PEM and DMFCs, while the regeneration process is difficult and ineffective. Both catalyst poisons, S and CO, originate from the anode's fuel (hydrogen) that is commonly produced from S to CO containing hydrocarbons, and the ppm-level traces of these pollutants remain even after the desulfurization and CO-removal processes are applied. The fuel cleanup operations are easily achievable for commodity hydrogen; however, these procedures become more complicated and expensive when they are performed as part of a man-portable system with an on board fuel reformer to allow use of liquid hydrocarbon fuels.

DMFCs are listed among the most promising technologies for smaller portable devices. The major advantage of DMFCs over PEM cells is the use of a liquid fuel, typically the mixture of methanol (20 vol%) and water (80 vol%). The use of liquid fuel solves the issue of the inefficiency of hydrogen storage. However, precious metal catalysts are still used in methanol fuel cells. The remaining major issue is CO poisoning of the catalysts as the fuel is reduced. Ruthenium (Ru)-based catalysts, commonly used in methanol fuel cells, provide a higher CO tolerance than Pt-based catalysts, but much higher level of CO tolerance is desired for commercial applications. There are many current research programs on methanol fuel cell catalysts that focus on producing more CO-tolerant catalysts. An additional disadvantage of DMFCs is the fact that the water-diluted methanol fuel is far from the chemical stoichiometric ratio required for oxidation of carbon to carbon dioxide, as seen in Eq. (7.4) that illustrates the anode half-reaction in a DMFC. The DMFC cathode half-reaction is analogous to that shown for a "classic" PEM cell in Eq. (7.3).

$$\text{Anode, DMFC:}\quad CH_3OH + H_2O \rightarrow 6H^+ + CO_2 \tag{7.4}$$

In addition to the electro-osmotic drag, methanol fuel cells suffer from significant crossover of the methanol from the anode to the cathode, which results in fuel consumption inefficiencies, which manifest as lowered energy and power densities. The two principal issues with DMFCs that prevent their broader market penetration are CO poisoning of the anode catalysts and the necessity for dilution of the fuel to mitigate crossover and electro-osmotic drag.

Table 7.1 Generic cell stack metrics for 150 W system

System and cell stack metrics	Generic performance
System power	150 W
Cell stack voltage	6 V
Cell voltage	0.6 V
Number of cells	10
Cell active area (membrane)	100 cm^2 (10 cm × 10 cm cell)
Current density	0.25 A cm^{-2}

PEM fuel cells are commercially available today in several niche applications. The cost associated with manufacturing the fuel cells and the need for compressed hydrogen storage mean that most units must typically exceed 100 W power output to be commercially viable. Several portable PEM units from Jadoo Power Systems for the military applications that deliver 150 W or more, are a good example of the minimum size limitation. Typically, lower power systems have a number of issues including Balance of Plant (BOP) complications. Most BOP electronics require 5 V of input voltage. This equates to a certain number of cells stacked in series within the cell stack to yield a voltage greater than 5 V. With an output voltage of the cell stack >5 V, a large range commercial-off-the-shelf (COTS) economical electronics can be used for BOP operations. Referencing a generic I–V (current-voltage) curve for a fuel cell gives an operating potential of 6 V per cell stack. The system and stack metrics can then be back calculated to give the generic metrics shown in Table 7.1. Note that balance-of-plant power requirements were ignored.

Examination of the cell stack metrics suggests several business and engineering limitations based on the assumption that the power and cell voltage are held constant. The business limitations relate to the current density being conservatively low in the developmental prototypes, whereas up to four times (1 A cm^{-2}) could be achieved in a commercial setting. This means that four times more raw materials are being used than necessary. However, from an engineering perspective, shrinking the cell to 25 cm^2 to increase the current density to 1 A cm^{-2} does not decrease the area around the edge of the active area available for sealing the cell. Generally speaking, a 1 cm-wide perimeter (excess) strip is still needed for sealing. This means 7 cm × 7 cm (49 cm^2) of a total membrane area to allow for sealing of only 25 cm^2 of the active membrane area. To minimize raw material costs, the sealing area needs to be a small fraction of the active area, not close to 50 %. Therefore, from a top-level system cost perspective, defined by cost in $\$\cdot W^{-1}$, little is gained by trying to engineer a system to less then 150 W.

The most persistent, unresolved issue with PEM fuel cells has always been the hydrogen storage. Despite advances in high pressure composite tanks, metal hydrides and metal organic frameworks, the energy density (J L^{-1}) of hydrogen is five times less than liquid fuels. Note that the specific energy is high for hydrogen but does not translate to volumetric storage, Table 7.2. This has led to PEM fuel cells being paired to fuel reformers so that hydrogen is generated on demand.

Table 7.2 Specific energy and energy density for various fuel types

Fuel type	HHV, MJ kg^{-1}	LHV, MJ kg^{-1}	Molecular mass g mol^{-1}	Density, kg L^{-1}	Weight, kg	Energy density, MJ L^{-1}
Hydrogen	141.86	–	2.02	–	0.06	7.97
Methane	55.53	–	16.04	–	0.45	24.77
Propane	–	46.44	–	0.493	0.49	22.89
Gasoline	–	44.50	–	0.710	0.71	31.60
JP-8	–	42.80	–	0.832	0.83	35.61
Diesel	–	42.50	–	0.832	0.83	35.36
Methanol	–	18.05	–	0.792	0.79	14.29

Calculation assumptions
Fuel volume is 1 L
Gaseous fuel compressed to 10,000 psi equals 27.81 mol of fuel
HHV used for gaseous fuels where no vaporization is required
LHV used for liquid fuels that require vaporization
Propane is used as a gas but the can be stored as a liquid above 2.2 MPa

A major funding source for the research on PEM fuel cells and on the associated fuel reformers has been the U.S. military.

The U.S. military has continually funded fuel cell R&D for longer term missions, typically 72 h. At this mission length, fuel cells are an attractive value proposition versus the BA-5290 military lithium-ion battery. Previous programs include Palm Power, Robust Portable Power, and Tactical Advanced Power (TAP) programs [1]. The military applications are open to all fuel cell technologies provided energy density and power density metrics are met, with the caveat that ultimately the technology should run off JP-8 (military diesel fuel). Currently, the U.S. military has a policy of fuel standardization on the battlefield, following the DOD Directive 4140.43. PEM cells coupled with reforming technology were initially tested on these programs but the need to deal with S and CO catalysts' poisons coming from the fuel has seen these programs gravitate toward high temperature Solid Oxide Fuel Cells (SOFC).

SOFCs have been developed for several man-portable applications. Typically, SOFCs operate at 800–1,000 °C although new materials have shown potential to operate at temperatures as low as 400 °C. To date, most of the SOFC systems have utilized propane as a fuel source due to its superior energy density at low pressure (2.2 MPa) compared to other gaseous fuels, as seen in Table 7.1. In an SOFC unit built with materials having standard insulation coefficients, the heat generated by electrical resistance is sufficient to preheat the inlet fuel streams while maintaining the operational temperature. The key advantage of SOFCs in these applications is the ability to provide heat for fuel reforming and being tolerant to CO poisoning of the anode catalysts. The CO poisoning of the catalyst occurs by absorption of the CO onto the catalyst surface. The surface adsorption of CO is temperature dependent and is significantly suppressed above 200 °C [2]. An additional advantage of SOFCs is that their ionic conduction mechanism is predominately based on O^- ionic transfer, which allows CO to be utilized as a fuel, as seen in the SOFC anode half-reaction, Eq. (7.5).

$$\text{Anode, SOFC:} \quad CO_{(g)} + 2O^{-} \rightarrow CO_{2(g)} \tag{7.5}$$

At temperatures above 780 °C, the following reaction occurs in addition to that presented in Eq. (7.5):

$$\text{Anode, SOFC}\left(>780\ ^{\circ}\text{C}\right): \quad CO_{2(g)} + C \rightarrow 2CO_{(g)} \tag{7.6}$$

In theory, an interpretation of Eq. (7.6) leads to the conclusion that the anode exhaust can be used to clean up soot deposits from SOFCs. However, the kinetics of this reaction are sluggish at best and, in most applications, a decrease in the operating temperature is sought since it prolongs the life of interconnected materials and helps to simplify other BOP equipment. The practical solution to soot formation from hydrocarbon fuel decomposition has been a partial oxidation reforming of the fuel stream. As with all technologies, there are drawbacks of the reforming approach. The reader is directed to J. Larminie, A. Dicks, "Fuel Cells Systems Explained", which offers good explanations of SOFC at the system level including the complications of fuel reforming.

A summary of the key advantages for man-portable applications of SOFCs are listed below:

(1) SOFCs can operate on liquid hydrocarbon fuels.
(2) The catalysts used in SOFCs are not Pt group metals.
(3) CO poisoning of the anode catalyst in an SOFC is not an issue.
(4) High-temperature operation helps with fuel reforming of hydrocarbon fuels.

There are several disadvantages of SOFCs which in some applications offset their desirable performance characteristics. The disadvantages are summarized below:

(1) SOFCs are predominantly made from ceramics which are brittle materials. The design of a man-portable system has to allow for the fact that these materials need to withstand shock from the end user actions (i.e., dropping the unit or subjecting it to vibrational forces during movement). Several companies, such as Protonex and Ultra Electronics AMI, have demonstrated robust system designs that accommodate the nonelastic properties of ceramic materials, although this does add a layer of complication to the system design.
(2) The catalysts are not sulfur tolerant. The majority of fossil-based liquid hydrocarbon fuels have significant sulfur content. Military diesel has the highest catalyst poison levels, with an allowable sulfur content of up to 3,000 ppm. Although fuel cleanup and sulfur removal can be achieved, they add significant system complexity and cost.
(3) High temperature complicates BOP components and requires significant use of exotic metal alloys such as nickel-based super alloys. These alloys have very high raw material costs and very high fabrication costs.
(4) The ceramic oxides used for the anode, electrolyte, and cathode in SOFCs are highly engineered and expensive. Although the predominant material for the electrolyte is zirconia (Zr_2O_3), commonly found in beach sand, the additives are rare earths including yttria (Y_2O_3) and scandia (Sc_2O_3). The cathode

electrode material typically is lanthanum strontium manganite (LSM) which also contributes a nontrivial portion to the system cost.

During the selection of a prospective energy generation solution for a man-potable application, both the power generation as well as the fuel storage must be taken into account. For short-term lower-power applications, such as smart phones, it is difficult to see an alternative to battery technology when the state-of-the-art metrics are considered. It should be noted that lithium-ion battery manufacturers are still making significant R&D progress. For long missions with higher power and energy demands, i.e., field deployable battery chargers (500–1,000 W for 72 h) fuel cells and sterling engines offer significant potential benefits over lithium-ion battery technology. Several highly engineered solutions have been developed and proven to be technically feasible; however, cost per power and cost per energy metrics are higher than permissible in the consumer market. The technical advantages of SOFCs provide enough reason for the U.S. military and other federal entities to continue the R&D support and promotion of these technologies.

7.2.3 Photovoltaics for Portable (and Other) Applications

Photovoltaic devices (vernacular–solar power) consist of a thin layer of the p-type conductor on top of a thin layer of n-type conductor to form a cell. The nomenclature of the structure is a p-n junction. For a more detailed explanation of semiconducting materials, the reader is directed to "Semiconductor Device Fundamentals" by Robert F. Pierret, listed in the reference section. Solar irradiation of certain semiconductor materials results in the breaking of the electron bonds which allows them to move freely in the p-type conductor. The n-type layer is negative with an excess of electrons while the p-type conductor with holes is the positive layer. The semiconductor materials are doped to control the electronic properties, Fig. 7.3. Currently, silicon (Si), gallium arsenide (GaAs), and copper sulphate (Cu_2S) are the materials of choice. Photovoltaics can be engineered to work very effectively in certain applications. For example, satellites and space probes have utilized arrays of photovoltaics for power generation. In 1958, Vanguard 1 was the first satellite launched with a photovoltaic array; modern examples include the Mars Global Surveyor. In these applications, the price of the array is of less importance than the functionality, with the material of choice being the more efficient (up to 29 %) but more expensive Si. The terrestrial issues associated with photovoltaics are different from the space applications' challenges, as there is no cloud cover in space. The terrestrial challenges can be mitigated with an engineering solution such as a suitable alignment of the array. At the other end of the spectrum, there are mass consumer products, such as some modern calculators that use a small array of photovoltaic cells to recharge the battery. This a convenient solution since calculators do not require much power and the array can use either office lighting or direct sunlight. However, currently there are some major barriers to mass commercial adoption of photovoltaics:

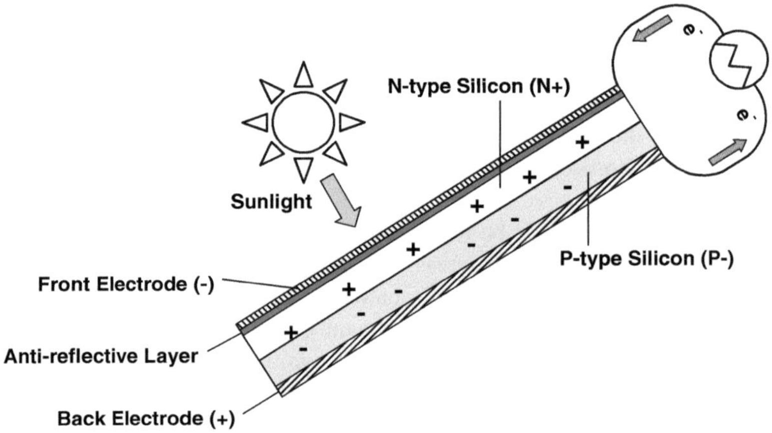

Fig. 7.3 Schematic of an electrical circuit within a photovoltaic device

(1) Cost: currently, the 3–7 kW array of photovoltaics for residential use costs \$18,000–\$40,000 [3]. Some sources [4] argue the cost of photovoltaics in the U.S. is unduly increased by the excessive federal regulations.
(2) Geographical Limitations: a photovoltaic array needs to be aligned to the sun and to be exposed to direct sunlight; shading from other structures and inclement weather conditions will limit effectiveness of the array.
(3) Efficiency: Photovoltaic cell's efficiency is dependent on the incident light that is available for conversion into electrical energy and can be reduced by such factors as dust accumulation or an obstacle in the path of sunlight.

There are niche applications for photovoltaic arrays across all the application categories: portable—personal calculators, transport—space probes, stationary—houses.

7.2.4 Thermal Electric Technology for Portable (and Other) Applications

Seebeck effect occurs from the application of a heat differential across a material that results in charge-carrier diffusion and phonon drag, Fig. 7.4. Traditionally, metals were used; however, today's state-of-the-art thermal electric devices use a combination of n-type and p-type semiconductor materials. Any heat source can be used, including radioisotopes, to generate a hot side for the device, which causes migration of the electrons in the n-type semiconductor to the cold side of the device. The flow of current is directionally proportional to holes' flow in the p-type semiconductor. Although thermal electric technology is well proven for electrical generation, thermal electric devices are inherently inefficient and require comparatively expensive materials for the energy output. As a result, the applications of thermal electric devices have mainly been limited to aerospace and military uses.

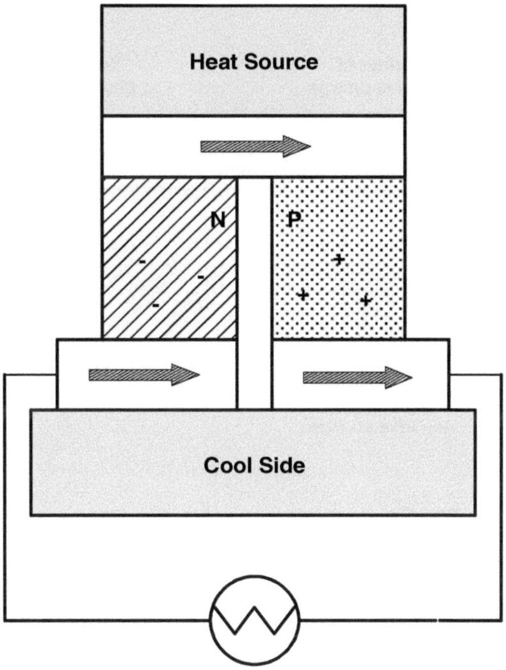

Fig. 7.4 Schematic of electrical circuit in a thermal electric device

For example, radioisotope thermal electric generators have been used on several space applications including the Mars rovers. Portable military heaters have also used thermal electric generators mainly for startup battery recharging.

7.2.5 Piezoelectric Technology for Portable Applications

The application of a compressive mechanical stress to certain crystals results in an electric charge referred to as a piezoelectric effect, Fig. 7.5. The best known such device is the gas stove starter that uses a piezoelectric to generate a very high voltage that causes the ignition spark. DARPA and others have investigated the use of piezoelectric devices as power generators. In DARPA's case, a boot-mounted device was tested to generate a couple of watts of power. In the case of the DARPA system, the extra energy expended by the soldiers and the comfort of the boots offset the benefits. This led to the DARPA program being canceled. However, for a very small portable device with ultra-low power requirements it is possible that a piezoelectric device could provide a reliable power source. The most likely scenario would be the hybrid of a piezoelectric unit and a battery that uses ambient vibrations to generate power that is stored in the battery. However, the power requirements of the majority of commercial devices are currently too demanding for the described system.

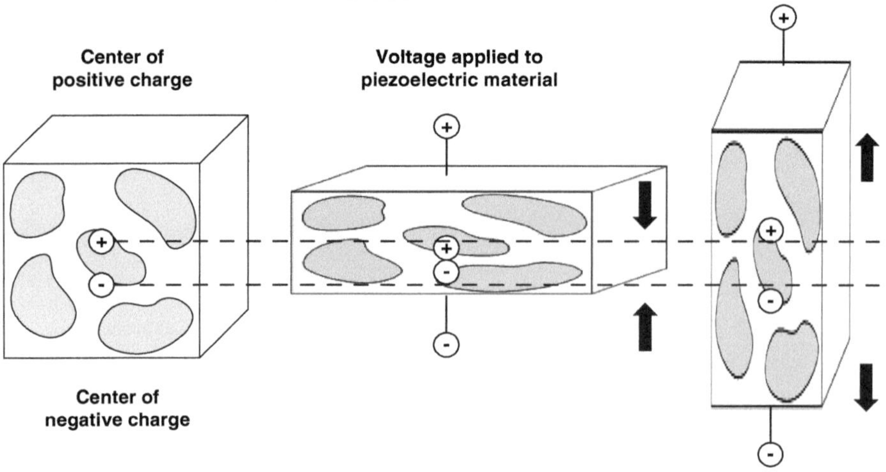

Fig. 7.5 Schematic of charge generation due to stress in a piezoelectric device

7.2.6 Supercapacitors for Portable Applications

Supercapacitors work due to the formation of Electrical Double Layers (EDL) on the interfaces of the electrolyte, Fig. 7.6. The double layers are formed when power is ap-plied to the device. Current can be drawn from the device once the layers are formed. However, the EDL have a limited lifetime and collapse over time back to an equilibrium state. The short lifetime of the EDL means that supercapacitors are limited to applications that have a short duty cycle. An automotive example would be storage of energy generated from regenerative braking for use in acceleration out of a corner or from a dead stop. It is likely that the supercapacitors will be used in an increasing range of applications but in conjunction with other energy storage and generation technologies.

7.2.7 Internal Combustion Engines for Portable Applications

At the upper limit of the man-portable power is the diesel generation set. Sold at the local hardware store for very affordable prices, typically a 1 kW generator costs less than $1/W. The reliability is unquestionable, with the unit handling many years of abuse in the field and with the end user being able to overhaul the unit easily and cost-effectively. In situations where diesel fuel is readily available at affordable prices, a diesel generator set will be selected over the alternative energy technologies. However, if the cost of fuel is high and other significant logistical issues occur, such as in the forward deployed military bases, then batteries, fuel cells, and other technologies will be more advantageous.

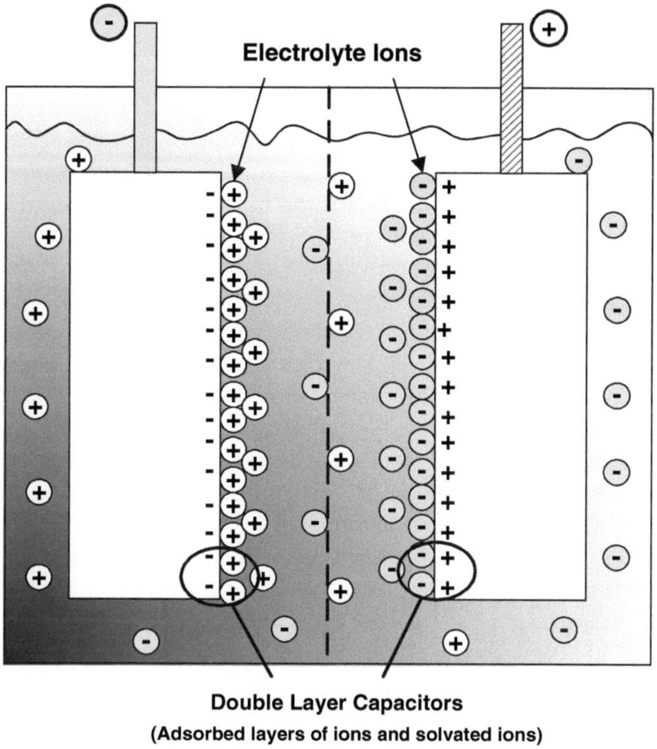

Fig. 7.6 Schematic of the electrical circuit within a supercapacitor

7.3 Transport Applications

Transport applications are currently dominated by the IC engine. Alternative technologies have to compete against two major facts. The first fact is that IC engines have been getting more efficient over time without a dramatic price premium. The second fact is that liquid fuels are the most efficient storage medium. In terms of battery technologies, lithium-ion batteries have and will continue to dominate. Development of new high energy cathode and anode materials will continue to keep lithium-ion chemistry as the dominant battery material for transport applications. However, the limited amount of storage in today's battery severely limits the range of the vehicles. It also takes considerable time to recharge the battery packs. The high costs associated with the battery materials make IC engines a clear winner for the foreseeable future. This is despite several automakers currently selling hybrid electric vehicles (HEV) and plug-in electric vehicles (PHEV).

Meanwhile, the design of IC engine has been evolving at a rapid pace. There are staggering differences between the 1960s IC engines compared to today's models. In fact, it is well understood by the automotive industry that the IC engine still has a lot of potential for improvement. This fact is illustrated by car companies agreeing to meet the new CAFÉ standards (mileage mandate) of 54.5 MPG for 2017–2025 models. Although there are technical challenges, the obstacles are mainly political and economical in nature. The cars are expected to cost more, approximately $3,000 per car [5]. Auto companies will have to run development programs to ensure the engineering aspects and costs are handled efficiently, but the technical solutions for the most part constitute of well understood, incremental changes. Therefore, the following question arises: "Is there anyone being highly innovative to provide the non-incremental, next-generation IC engines?" Despite the current lack of media attention, the answer is "yes".

In fact, there has been a number of venture capital-backed startup companies looking to develop a Gen 2 IC engine for the auto industry. Examples include Transonic Combustion, LiquidPiston, Eco Motors and Pinnacle Engines. LiquidPiston's product is based on a combination of several engine cycles (Otto, Diesel, Atkinson, and Rankine); however, in principle, this unit compresses and ignites a liquid hydrocarbon fuel. Initial testing results are promising for the above startups. Transonic Combustion dynamometer tests gave 98 MPG at steady state and 55 MPH. In practical terms, this means that the car would have approximately four times the range of today's latest model, which means the size of the fuel tank would shrink. Assuming a typical fuel tank of 70 L (18.5 gal), the fuel weight is approximately 50 kg, so a saving of 37.5 kg in weight and 75 % in volume can be realized by carrying less fuel.[5] The increase in fuel efficiency due to weight reduction for a typical car is insignificant (2 % at best); however, there is now space to fit another fuel-saving device onto the vehicle without adding weight. While the weight saving is not that significant, automakers are certainly not looking to add additional weight to the vehicle. Combine these innovative approaches with the incremental changes to the IC engine and it appears that the efficiency target for the alternative technologies is moving and could conceivably double within the next 10–20 years.

The one factor that may change the dominance of IC engines is a switch to another fuel source. The immediate pressure on a switch to alternative fuel sources is driven by the supply and price of oil, not by environmental concerns. In the short term, natural gas has been promoted as an alternative transport fuel. IC engines can be adapted to run on natural gas and other bio-liquid fuels. Although bio-liquid fuels have the ability to be carbon-neutral, currently this is not the case. Therefore, in the long term there is a need to use an alternative fuel that does not produce CO_2 tail pipe emissions for environmental reasons. Ultimately, this means a hydrogen-based economy. In the future automotive market, if the prevailing concept is that of a centralized hydrogen fuel source with electrical generation, then the electric car will be eventually be the winner. This is the most likely scenario if an energy-dense storage solution for hydrogen cannot be developed.

[5] Fuel density is assumed at 0.71 kg L^{-1} for this calculation; typically, this value ranges between 0.71–0.77 kg L^{-1}.

If a distributed hydrogen solution is successfully utilized, then the picture of the future car is less clear. Running IC engines off hydrogen is problematic but not im-possible. Therefore, fuel cells remain under active research and development by auto companies.

The automotive companies to date have focused on low-temperature PEM fuel cells, which originate from industry-dominated development efforts. Initially the effort was led by GE and later by United Technologies Center (UTC) whose Hamilton-Sundstrand division acquired GE fuel cell technology in the mid-1980s [6]. The R&D leadership in fuel cells is dominated by industry since the early development was performed in industry, not academia. This has resulted in industry holding the cumulative engineering knowledge. The engineering knowledge provides the ability to work with very thin membranes and ultra low platinum group metal (PGM) catalysts loading while maintaining the electrochemical system level performance. Since knowledge is "locked" within industrial trade secrets and has been rarely disclosed even in the form of patents, academic researchers are not fully aware of prior art.

All of the issues associated with fuel cells, previously mentioned for portable applications, hold true for transportable applications. In addition to the already known challenges, in the automotive application, the energy storage system has to regularly handle freezing temperature starting conditions. This is a major problem for a technology that has a membrane which requires tightly controlled humidity. Hence, the popular thought in the industry is that if automotive companies cannot commercialize PEM technology then it is unlikely to happen.

Other transportable applications also exist where fuel cells hold considerable advantages over batteries and other technologies. The U.S. air force has been developing SOFC fuel cells for full size Unmanned Aerial Systems (UAS). As with portable military applications, the mission metrics are in favor of fuel cells. The mission times are long, spanning at least 72 h, and the power requirements are reasonably constant. The ability of the SOFCs to provide heat for fuel reforming means they will continue to be the preferred fuel cell technology for this particular niche transport application. For smaller, mini and micro drones, the size, weight, and mission time requirements favor batteries or low temperature fuel cells. However, SOFCs have also been deployed on smaller UAS to increase mission time with Lockheed Martin's Stalker XE duration rated at 8+ h compared to the electrically propelled model, powered by a battery, that only has a 2+ h duration [7].

Other transport applications where fuel applications make sense include buses. Relatively speaking, buses have a large amount of space and a light load that enable on-board hydrogen storage. Another major advantage is that buses typically return to the depot to refuel which means that only a single refueling station is needed. UTC PEM fuel cells have been deployed on buses since 1988 in California and Connecticut [8].

The major focus of this section has been on describing the primary energy source for transport applications. There are, however, numerous transport applications where the auxiliary power is critical. The best recognized auxiliary power application is the automobile battery. The lead-acid chemistry has dominated this application since the lead-acid battery works under a wide range of operating temperatures and is inexpensive to manufacture and is reliable. Other energy storage and generation devices have also been utilized in a number of auxiliary power applications. In some cases, it is to improve efficiency through capture and release of energy. A good example of this is the Kinetic Energy Recovery System (KERS) installed on Formula 1 race cars, which uses a flywheel to capture energy under braking. The Zytek's Flybird system allows the cars to produce an additional 60 kW of energy for 6.6 s [9]. In other cases, operating off auxiliary power provides additional functionality to the device. There are several historical and modern military applications where the auxiliary power offers a distinct tactical advantage. World War II-era submarines were primarily powered by noisy diesel engines that required an air supply. Lead-acid batteries provided auxiliary power which allowed the submarines to submerge and run silent. In the modern military there is a range of equipment moved into place using conventional IC engine technology and then auxiliary power is used for silent monitoring of enemy locations. Batteries are currently deployed for these roles but longer mission times and increased power demands mean that the military is currently running development programs on batteries and fuel cells for these applications.

7.4 Stationary Applications

Stationary applications typically have very few weight and space limitations. For distributed power in residential settings there are typically some limitations; however, they are not as restrictive as the equivalent limits for portable or transportable applications. As the size of the energy storage device is proportional to the site of deployment, it is more likely to see an industrial application of the technology where weight and size are of secondary, or lesser, importance. The dominating metric is of course cost. The next differentiation is efficiency. Typically, at this scale, the efficiency relates to cost via power density. The greater the power density of the device, the lower the amount of raw materials required for construction of the unit. Typically, the cost of raw materials and not the construction cost begin to dominate as the scale of the device increases. Therefore, it is essential to consider the cost efficiency as a function of power density. Since it is typical to operate within a predefined voltage range, power density is therefore related to current density. In all electrochemical devices, the actual electrochemical potential is not constant but varies within a certain range; therefore, the current density that can be drawn (typically expressed in A cm^{-2}) is variable with the changing state of charge (SOC).

The cost of lithium-ion batteries is unfavorable compared to other devices that store energy. Since large-scale applications are driven by system cost, lithium-ion energy storage devices become impractical. In applications where greater than megawatt-level power is generated and stored, there is a host of other technologies that are more suitable. In terms of efficiency, any device that operates at a large scale and at a reasonable temperature, >300 °C, is advantageous since the heat can be used by the end user. Although lithium-ion batteries are routinely reported to be >90 % efficient, this result does not take into account the current density. Efficiency is a measure of the power generated by a given device as a function of the power consumed on the charge cycle. All electrochemical devices obey Ohm's law in which the potential is directly proportional to the current ($V = I \cdot R$). Since a constant potential is used to charge the battery, therefore the charge rate depends on the current. Assuming the battery is charged and discharged at the same rate then Ohm's law clearly states that the battery will be more efficient at low currents due to less electrical and ionic resistance. The effect of Ohm's law is seen on a discharge curve in any electrochemical energy storage device. At low C-rates, the middle section of the curve is almost a horizontal line. As the C-rate increases, the slope of the line becomes more negative and begins to mimic the $I \cdot V$ curve produced during a fuel cell testing. This tells the reader that the device with the lowest resistance is the most efficient. Thus, it is not surprising to find that most of the R&D efforts focused on improving the energy efficiency also focus on lowering the resistance. The major drawback of conventional batteries is the need to store the reactants in the electrodes. This prevents the energy storage section from being separated from the power generation device. At a large scale, the separation of energy and power becomes crucial.

The fundamental difference between a conventional battery and a fuel cell is the ability of the fuel cell's stack to be separate from the energy storage. This allows at a large scale for the power generation unit to be a modular system comprising of smaller units. Moreover, fuel cells are not unique in their ability to separate power from energy. For example, in a gas fired turbine power plant, additional turbines can be added or switched on to meet the current power demands. The energy storage segment is ideally a pipeline or modular tanks that are scaled to meet the demand. This provides the design team with the flexibility to control the cost of such system. It is not surprising then that there is a class of batteries and fuel cells that exhibit hybrid-like properties that are only applicable to large-scale stationary applications. Flow batteries, fuel cells with halogen-based chemistry, and sulfur batteries are common examples of such "hybrids".

One key aspect of these systems is the ability to run in reverse as energy storage devices. The introduction of renewable energy with variable power output into the grid can cause instability issues. In Germany and Hawaii, integrating of wind power into the grid at above 25 % caused the grid instability. To counteract this potential imbalance caused by renewable energy sources, systems that can store energy at the peak of generation and meter it back when required, are being developed for grid leveling applications. The most economical way of achieving this is the hydroelectric energy storage. During excess, power-generation pumps are run to lift water to the top reservoir. During high demand, water is released through the turbines at a cost of about 2 ¢/kWh. Geographical limitations render the solution impractical for widespread implementation.

Electrochemical devices for grid storage typically possess chemistries where the charge and discharge reactions have very similar kinetic limitations. An example would be the hydrogen-halogen flow batteries under development by Sustainable Innovations LLC, Lawrence Berkeley National Lab, and others. As with lithium-ion batteries, the efficiency of flow batteries is directly influenced by the drawn current density, which is dictated by Ohm's law. The other universal trend in electrochemical energy storage devices is their issue with scalability. The planar nature of most electrochemical energy storage technologies inherently leads to issues with controlling the fuel, temperature gradient, and current distribution throughout the device. As the footprint of the active area increases, the issues become more difficult to compensate for. This results in most electrochemical devices using multiple stacks to achieve the power requirements. Although there are companies that advocate filling large shipping containers with smaller batteries connected in packs, large-scale power generation for the grid requires scaling to size, and this scaling cannot be currently easily achieved for electrochemical energy storage devices.

An alternative to large-scale, centralized energy storage and generation is smaller, distributed power generation. In the long term, the introduction of more renewable energy sources and a smart grid implementation may facilitate the distributed power generation. However, currently fuel cells or batteries cannot produce and/or store the amount of energy required for stationary applications at reasonable cost.

7.5 Summary

Lithium-ion batteries compete against a number of emerging and established power generation and energy storage devices. In general, the metrics required of the devices are driven by whether the application is portable, transportable, or stationary.

In portable applications, secondary lithium-ion batteries dominate in the market for smaller, newer consumer products such as smart phones. Meanwhile, lithium-ion chemistry is being further adopted into the primary battery market. Due to lower costs, other battery chemistries such as zinc-air and zinc-carbon hold a significant market share and are likely to maintain their niche in the future. Larger non-consumer applications approaching 1 kW and 10 kWh have no well-defined power and energy storage solutions at the moment. Fuel cells meet the metrics in some applications but still have significant developmental issues. Lithium-ion chemistries are unlikely to dominate these applications unless radical improvements in energy density are achieved. Above 1 kW and 10 kWh, the IC engine dominates transportable applications.

Power and energy storage solutions for transportable applications typically gravitate toward the consumer automobiles. Currently, the IC engine dominates transport applications by delivering performance and reliability at a fraction of the cost of alternative technologies. New R&D efforts focused on improved IC engine efficiency to 2–4 times that of currently used commercial engines will generate further price pressure on alternative energy technologies. A switch to alternative non-hydrocarbon fuel sources such as hydrogen may finally displace the IC engine.

Lithium-ion batteries will emerge as the dominant power and energy source if the fuel is not carried on board the vehicles. If an energy dense storage solution for hydrogen is developed then it is likely that IC engines running on hydrogen will be used. Low-temperature PEM fuel cells and DMFCs will only emerge as the dominant technology if engineering solutions can be developed to solve their current limitations. Molten carbonate and solid oxide fuel cells that operate at high temperature are more suited to stationary applications.

In stationary applications that require power greater than 1 MW it is unlikely that lithium-ion batteries will ever hold a significant market share. In the majority of large stationary applications Combined Heat and Power (CHP) energy storage is highly efficient, with the modern combined cycle gas turbines currently being the technology of choice. The eventual transition away from hydrocarbon fossil fuels will most likely favor medium- and high-temperature fuel cells that can operate as CHP units. There are however significant engineering and cost-related hurdles for fuel cells to overcome. Photovoltaic technology is still not sufficiently cost-effective for mass power production and would need to be paired with an energy storage device. In the long term it is likely that stationary power generation will be achieved using a combined cycle gas turbine that uses hydrogen produced via renewable powered electrolysis. However, it is possible that lithium-ion battery technology may play a role in smaller, widely distributed stationary applications for grid stability purposes. A major reduction in price is required before this would be feasible.

7.6 Conclusions

A number of conclusions can be drawn about various energy storage technologies and their fit to various applications:

(1) At a small, portable scale, the performance per normalized weight and volume are the dominating metrics. At the large scale, it is efficiency and cost that prevail.
(2) The majority of energy storage technologies can be applied to a host of applications. However, typically one solution surpasses the others.
(3) The IC engine is the benchmark that alternative energy technologies must compete against, particularly in transport applications. IC engines are undergoing continued development with the possibility of a non-incremental change in performance from the second generation designs. This means that the IC engine technology as a benchmark is difficult to surpass.
(4) As research efforts on several alternative energy storage options reach developmental plateaus, hybridization of various technologies has been increasingly taking place to offer an acceptable system solution.

(5) Lithium-ion technology still has active R&D efforts. It is conceivable that lithium-ion technology will dominate the smaller higher power application field for the foreseeable future. Although the time line is unclear, it is likely that lithium-ion technology will be an energy and power source in future cars. A viable hydrogen storage technology will largely displace lithium-ion batteries but it is possible that they will still have a significant role in niche automobile markets.

7.7 Further Reading

Textbooks
Battery Technologies

(1) Reddy TB, Linden D, eds (2010) Linden's handbook of batteries 4th edition. McGraw-Hill, New York

Fuel Cells

(1) Larminie J, Dicks A (2003) Fuel cells systems explained 2nd edition. John Wiley & Sons, Ltd, Chichester, England

  – This reference covers the fundamental calculations for Nernst equation and other thermodynamic aspects of fuel cells; also provides an overview of system level design considerations.

(2) O'Hayre RP, Cha SW, Colella W, Printz FB (2006) Fuel cell fundamentals. John Wiley & Sons, Inc. Hoboken, New Jersey

  – Very good fundamental explanation of fuel cells with work examples.

(3) Vielstich W, Lamm A, Gasteiger HA (2009) Handbook of Fuel Cells, John Wiley & Sons, Ltd, Chichester, England

  – A fuel cells handbook, equivalent to the battery handbook. Note: this publication is a multiple (six) volume set, with a matching price tag.

(4) Hikosaka Behling N (2012) Fuel cells: current technology challenges and future research needs. Elsevier doi:10.1016/B978-0-444-56325-5.00005-3– Very comprehensive review of fuel cell technology.

Other Technologies

(1) Pierret RF (1996) Semiconductor device fundamentals, 2nd edition. Addison Wesley Publishing Company, Boston

  – Dated but nice fundamental textbook for semiconductor materials and devices.

(2) Markvart T, Castaner L (2003) Practical handbook of photovoltaics: fundamentals and Applications. Elsevier Science, Oxford, United Kingdom

(3) Winter M, Brodd R (2004) What are batteries, fuel cells, and supercapacitors. Chem Rev 104:4245–4270. doi:10.1021/cr020730k

- Excellent review paper for comparison of three different technologies. Deals with the scientific aspects as well the system level design.

(4) Arunachalam VS, Fleischer EL (2008) Harnessing materials for energy—review Issue. MRS Bull 33:1–255. doi:10.1557/mrs2008.60

- Broad range of topics is covered including oil, gas, hydro, biofuels as well as alternatives. This reference is focused predominantly on the materials research.

(5) Upbin B, ed (2008) Energy & genius. Forbes, Special Edition

- Mainly future fuel focused but relates this back to energy technologies and their respective economics.

(6) Meyers J (2010) Lightning in a bottle: storing energy for the smart grid. Interface, 9:44–63

- Overview of large scale battery technologies and capacitors for grid storage applications. Various flow batteries are also discussed.

Peer Reviewed Literature

General Fuel Cell Papers—to give the reader a board back ground introduction to various fuel cells types.

(1) Andújar JM, Segura F (2009) Fuel cells: history and updating. A walk along two centuries. Renew Sust Energ Rev, 13:2309–2322. doi:10.1016/j.rser.2009.03.015.
(2) Pollet BG, Staffell I, Shang JL (2012) Current status of hybrid, battery and fuel cell electric vehicles: from electrochemistry to market prospects. Electrochim Acta 84:235–249. doi:10.1016/j.electacta.2012.03.172
(3) Roche MY, et al. (2010) Public attitudes toward and demand for hydrogen and fuel cell vehicles: A review of the evidence and methodological implications. Energy Policy 38:5301–5310. doi:10.1016/j.enpol.2009.03.029
(4) Ang SMC, et al. (2011) Fuel cell systems optimisation—methods and strategies. Int J Hydrogen Energy 36:14678–14703. doi:10.1016/j.ijhydene.2011.08.053
(5) Borthwick WKD (2000) The European Union approach to fuel cell development. J Power Sources 86:52–56. doi:10.1016/S0378-7753(99)00462-0
(6) Mekhilef S, et al. (2012) Comparative study of different fuel cell technologies. Renew Sust Energ Rev 16:981–989. doi:10.1016/j.rser.2011.09.020
(7) Meibuhr SG Review of United States fuel-cell patents issued during 1965 and 1966. Electrochim Acta 13:1973 1995. doi:10.1016/0013-4686(68)80108-2
(8) Ogden JM, Steinbugler MM, Kreutz TG (1999) A comparison of hydrogen, methanol and gasoline as fuels for fuel cell vehicles: implications for vehicle design and infrastructure development. J Power Sources 79:143–168. doi:10.1016/S0378-7753(99)00057-9

(9) Acres GJK (2001) Recent advances in fuel cell technology and its applications. J Power Sources 100:60–66. doi:10.1016/S0378-7753(01)00883-7
(10) Oka Y (2009) Fuel Cells—Phosphoric acid fuel cells: cells and stacks, encyclopedia of electrochemical power sources 568–578. doi:10.1016/B978-044452745-5.00277-X
(11) Song C (2002) Fuel processing for low-temperature and high-temperature fuel cells: challenges, and opportunities for sustainable development in the 21st century. Catal Today 77:17–49. doi:10.1016/S0920-5861(02)00231-6
(12) Dicks AL (2006) The role of carbon in fuel cells. J Power Sources 156:128–141. doi:10.1016/j.jpowsour.2006.02.054
(13) Gahleitner G (2013) Hydrogen from renewable electricity: An international review of power-to-gas pilot plants for stationary applications. Int J Hydrogen Energy 38:2039–2061. doi:10.1016/j.ijhydene.2012.12.010
(14) Hemmes K, et al. (2012) Process intensification and fuel cells using a multi-source multi-product approach. Chem Eng Process: Process Intensification 51:88–108. doi:10.1016/j.cep.2011.09.010
(15) Mehta V, Cooper JS (2003) Review and analysis of PEM fuel cell design and manufacturing. J Power Sources. 114:32–53. doi:10.1016/S0378-7753(02)00542-6

Fuel Cell Economics and Market Analysis—a range of peer review literature that demonstrates the economic drivers for fuel cells within various market spaces

(1) Agnolucci P (2007) Economics and market prospects of portable fuel cells. Int J Hydrogen Energy 32:4319–4328. doi:10.1016/j.ijhydene.2007.03.042
(2) Jain S, Chen H, Schwank J (2006) Techno-economic analysis of fuel cell auxiliary power units as alternative to idling. J Power Sources 160:474–484. doi:10.1016/j.jpowsour.2006.01.083
(3) Staffell I, Green R (2013) The cost of domestic fuel cell micro-CHP systems. Int J Hydrogen Energy 38:1088–1102 doi:10.1016/j.ijhydene.2012.10.090
(4) Staffell I, Ingram A, Kendall K (2012) Energy and carbon payback times for solid oxide fuel cell based domestic CHP. Int J Hydrogen Energy 37:2509–2523. doi:10.1016/j.ijhydene.2011.10.060
(5) Ahluwalia RK, et al. (2004) Fuel economy of hydrogen fuel cell vehicles. J Power Sources 130:192–201. doi:10.1016/j.jpowsour.2003.12.061
(6) Wee J (2007) A feasibility study on direct methanol fuel cells for laptop computers based on a cost comparison with lithium-ion batteries. J Power Sources 173:424–436. doi:10.1016/j.jpowsour.2007.04.084
(7) Papadias DD, Ahmed S, Kumar R (2012) Fuel quality issues with biogas energy—An economic analysis for a stationary fuel cell system. Energy 44:257–277 doi:10.1016/j.energy.2012.06.031
(8) Schoots K, Kramer GJ, van der Zwaan BCC (2010) Technology learning for fuel cells: an assessment of past and potential cost reductions. Energy Policy 38:2887–2897. doi:10.1016/j.enpol.2010.01.022
(9) Hwang J, et al. (2013) Lifecycle performance assessment of fuel cell/battery electric vehicles. Int J Hydrogen Energy 38:3433–3446. doi:10.1016/j.ijhydene.2012.12.148

(10) Hawkes AD, Brett DJL, Brandon NP (2009) Fuel cell micro-CHP techno-economics: Part 1—model concept and formulation. Int J Hydrogen Energy 34:9545–9557. doi:10.1016/j.ijhydene.2009.09.094
(11) Hawkes AD, Brett DJL, Brandon NP (2009) Fuel cell micro-CHP techno-economics: Part 2—model application to consider the economic and environmental impact of stack degradation. Int J Hydrogen Energy 34:9558–9569. doi:10.1016/j.ijhydene.2009.09.095
(12) Cottrell CA, Grasman SE, Thomas M, Martin KB, Sheffield JW (2011) Strategies for stationary and portable fuel cell markets. Int J Hydrogen Energy 36:7969–7975. doi:10.1016/j.ijhydene.2011.01.056
(13) Budde B, Alkemade F, Weber KM (2012) Expectations as a key to understanding actor strategies in the field of fuel cell and hydrogen vehicles. Technological Forecasting and Social Change 79:1072–1083. doi:10.1016/j.techfore.2011.12.012
(14) Kang MJ, Park H (2011) Impact of experience on government policy toward acceptance of hydrogen fuel cell vehicles in Korea. Energy Policy 39:3465–3475. doi:10.1016/j.enpol.2011.03.045
(15) Sun Y, Ogden J, Delucchi M (2010) Societal lifetime cost of hydrogen fuel cell vehicles. Int J Hydrogen Energy 35:11932–11946. doi:10.1016/j.ijhydene.2010.08.044
(16) (2010) Comparative analysis of battery electric, hydrogen fuel cell and hybrid vehicles in a future sustainable road transport system. Energy Policy 38:24–29. doi:10.1016/j.enpol.2009.08.040
(17) Offer GJ, Contestabile M, Howey DA, Clague R, Brandon NP (2011) Techno-economic and behavioural analysis of battery electric, hydrogen fuel cell and hybrid vehicles in a future sustainable road transport system in the UK. Energy Policy 39:1939–1950. doi:10.1016/j.enpol.2011.01.006

Proton Exchange Membrane Fuel Cells—a selection of PEM fuel cell review paper on various key components and analysis techniques. High temperature PEM fuel cells are also included.

(1) Chandan A (2013) High temperature (HT) polymer electrolyte membrane fuel cells (PEMFC)—a review. J Power Sources 231:264–278. doi:10.1016/j.jpowsour.2012.11.126
(2) Wang Y, et al. (2011) A review of polymer electrolyte membrane fuel cells: Technology, applications, and needs on fundamental research. Appl Energ 88:981–1007. doi:10.1016/j.apenergy.2010.09.030
(3) Yousfi-Steiner N, Moçotéguy P, Candusso D, Hissel D (2009) A review on polymer electrolyte membrane fuel cell catalyst degradation and starvation issues: Causes, consequences and diagnostic for mitigation. J Power Sources 194:130–145. doi:10.1016/j.jpowsour.2009.03.060
(4) Dai W, et al. (2009) A review on water balance in the membrane electrode assembly of proton exchange membrane fuel cells. Int J Hydrogen Energy 34:9461–9478. doi:10.1016/j.ijhydene.2009.09.017
(5) Tawfik H, Hung Y, Mahajan D (2007) Metal bipolar plates for PEM fuel cell—a review. J Power Sources 163:755–767. doi:10.1016/j.jpowsour.2006.09.088

(6) Zhang S, et al. (2009) A review of accelerated stress tests of MEA durability in PEM fuel cells. Int J Hydrogen Energy 34:388–404. doi:10.1016/j.ijhydene.2008.10.012
(7) Yuan X, Wang H, Sun J C, Zhang J (2007) AC impedance technique in PEM fuel cell diagnosis—a review. Int J Hydrogen Energy 32:4365–4380. doi:10.1016/j.ijhydene.2007.05.036
(8) Cindrella L, et al. (2009) Gas diffusion layer for proton exchange membrane fuel cells—a review. J Power Sources 194:146–160. doi:10.1016/j.jpowsour.2009.04.005
(9) Wu J (2008) Diagnostic tools in PEM fuel cell research: Part I Electrochemical techniques. Int J Hydrogen Energy 33:1735–1746. doi:10.1016/j.ijhydene.2008.01.013
(10) Wu J (2008) Diagnostic tools in PEM fuel cell research: Part II: Physical/chemical methods. Int J Hydrogen Energy 33:747–1757. doi:10.1016/j.ijhydene.2008.01.020
(11) Zhang J, et al. (2006) High temperature PEM fuel cells. J Power Sources 160:872–891. doi:10.1016/j.jpowsour.2006.05.034
(12) Smitha B, Sridhar S, Khan AA (2005) Solid polymer electrolyte membranes for fuel cell applications—a review. J Membr Sci 259:10–26. doi:10.1016/j.memsci.2005.01.035

Direct Methanol/Ethanol Fuel Cells

(1) Kamaruddin MZF, Kamarudin SK, Daud WRW, Masdar MS (2013) An overview of fuel management in direct methanol fuel cells. Renew Sust Energ Rev 24:557–565. doi:10.1016/j.rser.2013.03.013
(2) Kamaruddin MZF, Kamarudin SK, Masdar MS, Daud WRW (2012) Review: direct ethanol fuel cells. Int J Hydrogen Energy. In Press. Corrected Proof. doi:10.1016/j.ijhydene.2012.07.059
(3) Basri S, Kamarudin SK (2011) Process system engineering in direct methanol fuel cell. Int J Hydrogen Energy 36:6219–6236. doi:10.1016/j.ijhydene.2011.02.058
(4) Scott K, Taama WM, Argyropoulos P (1999) Engineering aspects of the direct methanol fuel cell system. J Power Sources 79:43–59. doi:10.1016/S0378-7753(98)00198-0
(5) Kamarudin S K, Daud W R W, Ho S L, Hasran U A (2007) Overview on the challenges and developments of micro-direct methanol fuel cells (DMFC). J Power Sources 163:743–754. doi:10.1016/j.jpowsour.2006.09.081
(6) Kamarudin S K, Achmad F, Daud W R W (2009) Overview on the application of direct methanol fuel cell (DMFC) for portable electronic devices. Int J Hydrogen Energy 34:6902–6916. doi:10.1016/j.ijhydene.2009.06.013
(7) Dillon R, Srinivasan S, Aricò A S, Antonucci V (2004) International activities in DMFC R&D: status of technologies and potential applications. J Power Sources 127:112–126. doi:10.1016/j.jpowsour.2003.09.032
(8) S. Wasmus, A. Küver (1999) Methanol oxidation and direct methanol fuel cells: a selective review. J Electroanal Chem 461:14–31. doi:10.1016/S0022-0728(98)00197-1

(9) Liu H, et al. (2006) A review of anode catalysis in the direct methanol fuel cell. J Power Sources 155:95–110. doi:10.1016/j.jpowsour.2006.01.030
(10) Zainoodin A M, Kamarudin S K, Daud W R W (2010) Electrode in direct methanol fuel cells. Int J Hydrogen Energy 35:4606–4621. doi:10.1016/j.ijhydene.2010.02.036
(11) Li X, Faghri A (2013) Review and advances of direct methanol fuel cells (DMFCs) part I: Design, fabrication, and testing with high concentration methanol solutions. J Power Sources 226:223–240. doi:10.1016/j.jpowsour.2012.10.061

Alkaline Fuel Cells

(1) Bidault F, Brett DJL, Middleton PH, Brandon NP (2009) Review of gas diffusion cathodes for alkaline fuel cells. J Power Sources 187:39–48. doi:10.1016/j.jpowsour.2008.10.106
(2) Merle G, Wessling M, Nijmeijer K (2011) Anion exchange membranes for alkaline fuel cells: a review. J Membr Sci 377:1–35. doi:10.1016/j.memsci.2011.04.043
(3) McLean GF, Niet T, Prince-Richard S, Djilali N (2002) An assessment of alkaline fuel cell technology. Int J Hydrogen Energy 27:507–526. doi:10.1016/S0360-3199(01)00181-1
(4) Staffell I, Ingram A (2010) Life cycle assessment of an alkaline fuel cell CHP system. Int J Hydrogen Energy 35:2491–2505. doi:10.1016/j.ijhydene.2009.12.135
(5) Bidault F, Brett DJL, Middleton PH, Abson N, Brandon NP (2010) An improved cathode for alkaline fuel cells. Int J Hydrogen Energy 35:1783–1788. doi:10.1016/j.ijhydene.2009.12.035
(6) Qiao J, et al. (2013) Carbon-supported co-pyridine as non-platinum cathode catalyst for alkaline membrane fuel cells. Electrochim Acta 96:298–305. doi:10.1016/j.electacta.2013.02.030
(7) Zhou G, Chen L, Seaba JP (2011) CFD prediction of shunt currents present in alkaline fuel cells. J Power Sources 196:8180–8187. doi:10.1016/j.jpowsour.2011.04.029
(8) Gouérec P, Poletto L, Denizot J, Sanchez-Cortezon E, Miners JH (2004) The evolution of the performance of alkaline fuel cells with circulating electrolyte. J Power Sources 129:193–204. doi:10.1016/j.jpowsour.2003.11.032

Phosphoric Acid Fuel Cells

(1) Oka Y (2009) Fuel Cells—Phosphoric acid fuel cells: cells and stacks, encyclopedia of electrochemical power sources 568–578. doi:10.1016/B978-044452745-5.00277-X
(2) Kwak H, Lee H, Jung J, Jeon J, Park D (2004) Exergetic and thermoeconomic analysis of a 200-kW phosphoric acid fuel cell plant. Fuel 83:2087–2094. doi:10.1016/j.fuel.2004.04.002
(3) Hojo N, Okuda M, Nakamura M (1996) Phosphoric acid fuel cells in Japan, J Power Sources 61:73–77. doi:10.1016/S0378-7753(96)02339-7
(4) Ma Z, Venkataraman R, Farooque M (2009) Fuel cells—molten carbonate fuel cells modeling, encyclopedia of electrochemical power sources 519–532. doi:10.1016/B978-044452745-5.00272-0

Molten Carbonate Fuel Cells

(1) Giddey S, Badwal SPS, Kulkarni A, Munnings C (2012) A comprehensive review of direct carbon fuel cell technology. Prog Energ Combust 38:360–399. doi:10.1016/j.pecs.2012.01.003
(2) Yuh C, Johnsen R, Farooque M, Maru H (1995) Status of carbonate fuel cell materials. J Power Sources 56:1–10. doi:10.1016/0378-7753(95)80001-W
(3) Farooque M, Maru HC (2006) Carbonate fuel cells: milliwatts to megawatts. J Power Sources 160:827–834. doi:10.1016/j.jpowsour.2006.04.127
(4) Nguyen QM (1988) Technological status of nickel oxide cathodes in molten carbonate fuel cells—a review. J Power Sources 24:1–19. doi:10.1016/0378-7753(88)80085-5
(5) Antolini E (2011) The stability of molten carbonate fuel cell electrodes: a review of recent improvements. Appl Energ 88:4274–4293. doi:10.1016/j.apenergy.2011.07.009
(6) Wee J (2011) Molten carbonate fuel cell and gas turbine hybrid systems as distributed energy resources. Appl Energy 88:4252–4263. doi:10.1016/j.apenergy.2011.05.043
(7) Campanari S, Chiesa P, Manzolini G (2010) CO_2 capture from combined cycles integrated with molten carbonate fuel cells. Int. J. Greenhouse Gas Control 4:441–451. doi:10.1016/j.ijggc.2009.11.007
(8) Discepoli G, Cinti G, Desideri U, Penchini D, Proietti S (2012) Carbon capture with molten carbonate fuel cells: Experimental tests and fuel cell performance assessment. Int. J. Greenhouse Gas Control. 9:372–384. doi:10.1016/j.ijggc.2012.05.002
(9) Wee J, Roh JH, Kim J (2011) A comparison of solar photovoltaics and molten carbonate fuel cells as commercial power plants. Renew Sust Energ Rev 15:697–704. doi:10.1016/j.rser.2010.09.012
(10) Yuh C, Farooque M (2009) Fuel cells—molten carbonate fuel cells: materials and life considerations. Encyclopedia of Electrochemical Power Sources 497–507. doi:10.1016/B978-044452745-5.00270-7
(11) Farooque M, Maru H (2009) Fuel cells—molten carbonate fuel cells: full-scale prototypes. Encyclopedia of Electrochemical Power Sources 508–518. doi:10.1016/B978-044452745-5.00269-0
(12) Selman JR, Chen C (2012) Scientific and technical maturity of molten carbonate technology. Int J Hydrogen Energy 37:19280–19288. doi:10.1016/j.ijhydene.2012.06.016

Solid Oxide Fuel Cells

(1) Tucker MC (2010) Progress in metal-supported solid oxide fuel cells: a review. J Power Sources 195:4570–4582. doi:10.1016/j.jpowsour.2010.02.035
(2) Mahapatra MK, Lu K (2010) Seal glass for solid oxide fuel cells. J Power Sources 195:7129–7139. doi:10.1016/j.jpowsour.2010.06.003

(3) Wang K, et al. (2011) A review on solid oxide fuel cell models. Int J Hydrogen Energy 36:7212–7228. doi:10.1016/j.ijhydene.2011.03.051
(4) Stambouli A B, Traversa E (2002) Solid oxide fuel cells (SOFCs): a review of an environmentally clean and efficient source of energy. Renew Sustain Energ Rev 6:433–455. doi:10.1016/S1364-0321(02)00014-X
(5) Barelli L, Barluzzi E, Bidini G (2013) Diagnosis methodology and technique for solid oxide fuel cells: a review. Int J Hydrogen Energy 38:5060–5074. doi:10.1016/j.ijhydene.2013.02.024
(6) Singhal SC (2000) Advances in solid oxide fuel cell technology. Solid State Ionics 135:305–313. doi:10.1016/S0167-2738(00)00452-5
(7) Lawlor V, et al. (2009) Review of the micro-tubular solid oxide fuel cell: part I. Stack design issues and research activities. J Power Sources 193:387–399. doi:10.1016/j.jpowsour.2009.02.085
(8) Huang K, Singhal SC (2013) Cathode-supported tubular solid oxide fuel cell technology: a critical review 237:84–97. doi:10.1016/j.jpowsour.2013.03.001
(9) Mahapatra MK, Lu K (2010) Glass-based seals for solid oxide fuel and electrolyzer cells—a review. Mat Sci Eng R 67:65–85. doi:10.1016/j.mser.2009.12.002
(10) Steele BCH (1994) Oxygen transport and exchange in oxide ceramics. J Power Sources 49:1–14. doi:10.1016/0378-7753(93)01789-K
(11) Steele BCH (1996) Ceramic ion conducting membranes. Curr Opin Solid St M 1:684–691. doi:10.1016/S1359-0286(96)80052-0
(12) Steele BCH (1996) Survey of materials selection for ceramic fuel cells II. Cathodes and anodes. Solid State Ionics 86–88:1223–1234. doi:10.1016/0167-2738(96)00291-3
(13) Steele BCH (1993) Materials for electrochemical energy conversion and storage systems. Ceram Int 19:269–277. doi:10.1016/0272-8842(93)90059-Z
(14) Minh N Q (2004) Solid oxide fuel cell technology—features and applications. Solid State Ionics 174:271–277. doi:10.1016/j.ssi.2004.07.042
(15) Minh NQ (1993) Ceramic Fuel Cells. J Am Ceram Soc 76:563–588. doi:10.1111/j.1151-2916.1993.tb03645.x
(16) Wu J, Liu X (2010) Development of SOFC metallic interconnect. J Mater Sci Technol 26:293–305. doi:10.1016/S1005-0302(10)60049-7
(17) Zhang X, et al. (2010) A review of integration strategies for solid oxide fuel cells. J Power Sources 195:685–702. doi:10.1016/j.jpowsour.2009.07.045

Microbial Fuel Cells

(1) Osman MH, Shah AA, Walsh FC (2011) Recent progress and continuing challenges in bio-fuel cells. Part I: Enzymatic cells. Biosens Bioelectron 26:3087–3102. doi:10.1016/j.bios.2011.01.004
(2) Osman MH, Shah AA, Walsh FC (2010) Recent progress and continuing challenges in bio-fuel cells. Part II: Microbial. Biosens and Bioelectron 26:953–963. doi:10.1016/j.bios.2010.08.057

(3) Oliveira VB, Simões M, Melo LF, Pinto A M F R (2013) Overview on the developments of microbial fuel cells. Biochem Eng J 73:53–64. doi:10.1016/j.bej.2013.01.012
(4) Sundarrajan S, Allakhverdiev SI, Ramakrishna S (2012) Progress and perspectives in micro direct methanol fuel cell. Int J Hydrogen Energy 37:8765–8786. doi:10.1016/j.ijhydene.2011.12.017

Other Fuel Cells

(1) Ma J, Choudhury NA, Sahai Y (2010) A comprehensive review of direct borohydride fuel cells. Renew Sust Energ Rev 14:183–199. doi:10.1016/j.rser.2009.08.002
(2) Kjeang E, Djilali N, Sinton D (2009) Microfluidic fuel cells: a review. J Power Sources 186:353–369. doi:10.1016/j.jpowsour.2008.10.011
(3) Giddey S, Badwal SPS, Kulkarni A, Munnings C (2012) A comprehensive review of direct carbon fuel cell technology. Prog Energ Combust 38:360–399. doi:10.1016/j.pecs.2012.01.003
(4) Serov A, Kwak C (2010) Direct hydrazine fuel cells: a review. Appl Catal B: Environ 98:1–9. doi:10.1016/j.apcatb.2010.05.005
(5) Bartrom A M, et al. (2013) Optimization of an anode fabrication method for the alkaline Direct Formate Fuel Cell. J Power Sources 229:234–238. doi:10.1016/j.jpowsour.2012.12.007

Photovoltaics

(1) Raman P, Murali J, Sakthivadivel D, Vigneswaran VS (2012) Opportunities and challenges in setting up solar photo voltaic based micro grids for electrification in rural areas of India. Renew Sust Energ Rev 16:3320–3325. doi:10.1016/j.rser.2012.02.065
(2) Islam T, Meade N (2013) The impact of attribute preferences on adoption timing: The case of photo-voltaic (PV) solar cells for household electricity generation. Energy Policy 55:521–530. doi:10.1016/j.enpol.2012.12.041
(3) Bhattacharyya S C (2013) Financing energy access and off-grid electrification: a review of status, options and challenges. Renew Sust Energ Rev 20:462–472. doi:10.1016/j.rser.2012.12.008
(4) Mousazadeh H, et al. (2009) A review of principle and sun-tracking methods for maximizing solar systems output. Renew Sust Energ Rev 13:1800–1818. doi:10.1016/j.rser.2009.01.022
(5) Nema P, Nema RK, Rangnekar S (2009) A current and future state of art development of hybrid energy system using wind and PV-solar: a review. Renew Sust Energ Rev 13:2096–2103. doi:10.1016/j.rser.2008.10.006
(6) Davies P A, Luque A (1994) Solar thermophotovoltaics: brief review and a new look. Sol Energ Mat and Sol C 33:11–22. doi:10.1016/0927-0248(94)90284-4
(7) Bajpai P, Dash V (2012) Hybrid renewable energy systems for power generation in stand-alone applications: a review. Renew Sust Energ Rev 16:2926–2939. doi:10.1016/j.rser.2012.02.009

(8) Leijon M, Skoglund A, Waters R, Rehn A, Lindahl M (2010) On the physics of power, energy and economics of renewable electric energy sources—Part I. Renew Energ 35:1729–1734. doi:10.1016/j.renene.2009.10.030
(9) Skoglund A, Leijon M, Rehn A, Lindahl M, Waters R (2010) On the physics of power, energy and economics of renewable electric energy sources—Part II. Renew Energ 35:1735–1740. doi:10.1016/j.renene.2009.08.031
(10) Azadian F, Radzi MAM (2013) A general approach toward building integrated photovoltaic systems and its implementation barriers: a review. Renew Sust Energ Rev 22:527–538. doi:10.1016/j.rser.2013.01.056

Other Technologies

(1) Scrosati B (2005) Power Sources for Portable Electronics and Hybrid Cars: Lithium Batteries and Fuel Cells. Chem Rec 5:286–297. doi:10.1002/tcr.20054– Review article on fuel cell and lithium-ion batteries for non-stationary applications.
(2) Tanaka Y (2006) Advanced Metals and Ceramics for Clean Energy Technologies. IEE J Trans Fund Mater 126:24–25. doi:10.1541/ieejfms.126.24– Material perspective on a number of difference technologies.
(3) Kongtragool B, Wongwises S (2003) A review of solar-powered Stirling engines and low temperature differential Stirling engines. Renew Sust Energ Rev 7:131–154. doi:10.1016/S1364-0321(02)00053-9
(4) Thombare DG, Verma SK (2008) Technological development in the Stirling cycle engines. Renew Sust Energ Rev 12:1–38. doi:10.1016/j.rser.2006.07.001

References

1. http://www.darpa.mil, DARPA TAP Program. September 2, 2011 news release
2. Scrosati B (2005) Power Sources for Portable Electronics and Hybrid Cars: Lithium Batteries and Fuel Cells, Chem. Rec., 5:286–297 doi: 10.1002/tcr.20054
3. http://www.sunrunhome.com/solar-lease/cost-of-solar/. Accessed Jan 2013
4. Upbin B, ed (2008) Energy & genius. Forbes, Special Edition
5. Geman B, Laing K, Obama finalizes auto mileage mandate. E2 Wire, Accessed August 28 2012
6. Hikosaka Behling N (2012) Fuel Cells: Current Technology Challenges and Future Research Needs, Elsevier. doi:10.1016/B978-0-444-56325-5.00005-3
7. http://www.lockheedmartin.com/us/products/stalker-uas.html. Accessed Jan 2013
8. http://www.utcpower.com. Accessed Jan 2013
9. http://www.racecar-engineering.com. Accessed Jan 2013